Preface

Welcome to Internet for the Molecular Biologist. The object of this book is a very simple one: to introduce the Internet to that section of molecular biologists who do not consider themselves to be "computer literate". Until fairly recently the world wide traffic of information via computer interconnections was really the domain of the initiated, in most cases an in-depth understanding of the computers themselves was required and the preparation of material for distribution was not trivial. However, as the Internet expands the wealth of information available increases but the means by which that information can be accessed are becoming greater and simpler to use. The Internet has now developed into a truly useful tool and, like most great tools, one doesn't need to know how it works, just how to use it. A deep understanding of the machines, protocols and connections is no longer necessary. That said, it is often easier to use a tool if you have an idea of how it works. So the object of this book is to impart an understanding of the Internet without shrouding it in technical information. The early parts of the book should give the reader an insight into the principles of the Internet's generation, its evolution and its uses.

This is followed by consideration of some key information sources that should be of use to a wide range of scientists. Finally there are several chapters giving brief descriptions of a selection of information sources that should be of interest to particular areas of molecular biology.

This book should serve as an example of the versatility of the Internet. The entire volume was produced from concept to submission by using the Internet. The work was proposed by the publishers using the bionet newsgroups. The editors replied to the publishers using Email and communicated with each other exclusively by Email. Contributing authors were contacted by Email and information was pooled between these people using a mailing list. Authors submitted their chapters to the editors either via Email or by ftp. Final copy was then delivered to the publishers by ftp. To complete the picture a WWW page associated with this work has also been constructed and can be accessed at the following URL:

`http://www.ccc.nottingham.ac.uk/~mbzsrs/IFTMB.HTML`

If there is anything in the above paragraph that you don't understand don't worry. Read the book and you should!

The editors wish to express their thanks to everyone involved in the production of this book and in particular the contributing authors. Thank you.

Simon Swindell, Research Fellow, Special Projects Group, Department of Biochemistry, Queen's Medical Centre, Nottingham University, Nottingham, NG7 2UH. UK.
Email: Simon.Swindell@Nottingham.ac.uk
http://www.ccc.nottingham.ac.uk/~mbzsrs/Home.HTML

R. Russell Miller, Director of Technology Applications Development Office of Technology Transfer Lockheed Martin Energy Systems, Inc. (Managing Contractor for the U.S Department of Energy facilities in Oak Ridge, Tennessee)
Email: r69@ornl.gov

Garry S. A. Myers, PhD Student, Molecular Genetics Unit, Menzies School of Health Research, Darwin, Northern Territory, Australia.
Email: garry@menzies.su.edu.au

The *Current Innovations in Molecular Biology* series

Volume		Publication Date
1.	Molecular Biology: Current Innovations and Future Trends. Part 1.	April 1995
2.	Molecular Biology: Current Innovations and Future Trends. Part 2.	October 1995
3.	Internet for the Molecular Biologist.	January 1996

Coming soon:
Genetic Engineering with PCR, Robert M. Horton (Ed.).
Gene Cloning and Analysis: Current Innovations, Brian Schaefer (Ed.).

For further information on these books please contact:

USA/North America

Horizon Scientific Press
ISBS
5804 NE Hassalo Street
Portland
Oregon 97213-3644

Tel: (503) 287-3093
or (800) 944-6190
Fax: (503) 280-8832

Rest of World

Horizon Scientific Press
P.O. Box 1
Wymondham
Norfolk
NR18 0EH
U.K.

Fax: 0953-603068
International fax: +44-1953-603068

http://www.apollo.co.uk/a/horizon

INTERNET
FOR THE
MOLECULAR
BIOLOGIST

Volume 3 in the *Current Innovations in Molecular Biology* series

Edited by

Simon R. Swindell
Department of Biochemistry, Queen's Medical Centre, Nottingham University, Nottingham, NG7 2UH, England

R. Russell Miller
Lockheed Martin Energy Systems, Inc., Oak Ridge, Tennessee, USA

Garry S. A. Myers
Molecular Genetics Unit, Menzies School of Health Research, Darwin, Northern Territory, Australia

 **horizon scientific press**

Copyright © 1996
Horizon Scientific Press
P.O. Box 1
Wymondham
Norfolk NR18 0EH
England
http://www.apollo.co.uk/a/horizon

British Library Cataloguing-in-Publication Data

Internet for the Molecular Biologist. -
(Current Innovations in Molecular Biology)
    I. Swindell, S. R. II. Series
    574.88028546

ISBN: 1-898486-02-6

Printed in England by the Ipswich Book Company, Suffolk
Reprinted 1997

USA/North America

Horizon Scientific Press
ISBS
5804 NE Hassalo Street
Portland
Oregon 97213-3644

Tel: (503) 287-3093 or (800) 944-6190
Fax: (503) 280-8832

Australia/New Zealand

DA Direct
648 Whitehorse Road
Mitcham
3132 Australia

Tel. (03) 873 4411
Fax. (03) 873 5679
http://www.dadirect.com.au

Contents

Contributors

John Barnett
Genome Informatics Group
National Agricultural Library
Agricultural Research Service, USDA
Beltsville, MD 20705-2351

Stephen M. Beckstrom-Sternberg
Genome Informatics Group
National Agricultural Library
Agricultural Research Service, USDA
Beltsville, MD 20705-2351

Doug Bigwood
Genome Informatics Group
National Agricultural Library
Agricultural Research Service, USDA
Beltsville, MD 20705-2351

Sam Cartinhour
Crop Biotechnology Center
Texas A&M University
College Station, TX 77843

James W. Fickett
Theoretical Biology and Biophysics
Group
MS K710 Los Alamos National
Laboratory
Los Alamos, NM 87545

Gordon Findlay
Christchurch School of Medicine
Christchurch
New Zealand.

Gustavo Glusman
Department of Membrane Research &
Biophysics
Weizmann Institute of Science
Rehovot 76100
Israel

Roderic Guigó
Departament d'Informàtica Mèdica
Institut Municipal d'Investigació Mèdica
C/ Doctor Aiguader 80,
08003 Barcelona, Spain

Stephen Heller
Plant Genome Program
Building 005, BARC-W
Agricultural Research Service, USDA
Beltsville, MD 20705-2350

Kathie T. Hodge
Dept. of Plant Pathology
Cornell University
Ithaca, NY 14853
USA

David Hodgson
The Sanger Centre
Hinxton Hall
Cambridge
CB10 1RQ

Steven J.M. Jones
The Sanger Centre
Hinxton Hall
Cambridge
CB10 1RQ

Gail Juvik
Genome Informatics Group
National Agricultural Library
Agricultural Research Service, USDA
Beltsville, MD 20705-2351

M.A.Kennedy
Cytogenetic and Molecular Oncology
Unit
Christchurch School of Medicine
New Zealand

10

Jon Krainak
Genome Informatics Group
National Agricultural Library
Agricultural Research Service, USDA
Beltsville, MD 20705-2351

David Kristofferson, Ph.D., M.B.A.
Director of Electronic Services
IntelliGenetics, Inc.
700 E. El Camino Real, Suite 300
Mountain View, CA 94040

Martin Latterich
Molecular Biology and Virology
Laboratory
The Salk Institute
10010 N. Torrey Pines Road
La Jolla, California 92037

Jill Martin
Genome Informatics Group
National Agricultural Library
Agricultural Research Service, USDA
Beltsville, MD 20705-2351

Eric Mercer
Department of Membrane Research &
Biophysics
Weizmann Institute of Science
Rehovot 76100
Israel

Jerome P. Miksche
Plant Genome Program
Building 005, BARC-W
Agricultural Research Service, USDA
Beltsville, MD 20705-2350

Tahir S. Pillay, MD, PhD
Division of Endocrinology &
Metabolism
University of California, San Diego
9500 Gilman Drive, La Jolla
CA 92093-0673

Irit Rubin
Department of Membrane Research &
Biophysics
Weizmann Institute of Science
Rehovot 76100
Israel

Michael Shives
Genome Informatics Group
National Agricultural Library
Agricultural Research Service, USDA
Beltsville, MD 20705-2351

Marty Sikes
Genome Informatics Group
National Agricultural Library
Agricultural Research Service, USDA
Beltsville, MD 20705-2351

From: *Internet for the Molecular Biologist.*
ISBN 1-898486-02-6 ©1996 Horizon Scientific Press, Wymondham, U.K.

1

THE INTERNET:
WHAT, WHERE, WHO, WHEN

Gordon Findlay

Introduction

The purpose of this chapter is to introduce the Internet, briefly describe its history and governance and show some of the ways in which it can be used for serious scientific purposes.

What is it?

It doesn't really help to talk in depth about the technicalities which allow users of a computer in Christchurch, New Zealand to send messages to users in Finland and Singapore; to get a PC game from Massachusetts; to ask a question and get answers to it from Bath and San Francisco within hours; to perform an homology search against a database in Japan, and to search for a reference in library catalogues in Australia and Canada, all in a quiet morning. Nevertheless the question keeps coming up: what is the Internet?

The Internet is a loosely organised, interconnected set of computer networks that share a communication protocol (Transmission Control Protocol/Internet Protocol, or TCP/IP). TCP/IP allows a wide variety of computers to "talk" to each other in a common format. These interconnected networks have agreed on how to use this protocol to perform some common tasks. Computers on any of the networks are known as a node. To use the Internet you must gain access to (or own!) a node.

That's the physical view of the Internet - interconnected computers. From another viewpoint, the Internet is an interconnected network of services, which may be physically located in a particular place, but which are available for users world wide. The scientist performing a BLAST search needs to know about how to specify a search, how to enter a sequence and an address to send the search to. It isn't necessary to know what sort of computer the search will run on, nor where the computer is; there is a service "somewhere".

Yet another view of the Internet is that of a community (or number of communities) defined by common interests. Whereas communities have, in the past, been defined mainly in terms of location, in cyberspace communities define themselves in terms of interests and goals. The essence of a community is communication, and the Internet frees the channels of communication from most of the physical limitations that previously restricted community growth.

Where Did it Come From?

The roots of the Internet lie in earlier networking experiments such as ARPANET (the Advanced Research Projects Agency Network) dating from the late 1960s. The experiments that constitute the very start of ARPANET were trying to devise a method of distributing data (and people) so that a chain of command could still be maintained after a nuclear strike. An experimental network based on packet-switching technology was set up, with two revolutionary new features. Multiple users could share a communication line, and routing was invented so that communication was possible between computers which were not directly linked.

Electronic mail may not have been high in the researchers' thinking, but soon took off. The researchers themselves, and soon other scientists, discovered the benefits of being able to send detailed messages at something approaching the speed of a telephone call. Various other networks were set up in response to the need of academics and researchers to communicate with each other, and to share resources. The most important of these networks were BITnet CSnet and the NSFNET, commissioned by the National Science Foundation in the United States. ARPA was again instrumental in ensuring the development of the protocols which make it possible for various machines and networks to inter-communicate.

These networks have gradually coalesced, agreed on one naming scheme (although there are vestiges of older schemes still) and become known as the Internet, with their individual identities submerged. The whole process has been an organic one; more of the history can be found in the article by Hardy mentioned in the references. Until the late 1980s the net was primarily the domain of academic institutions and government research agencies. Private research companies began to join the net in the late 1980s, followed by commercial companies. The net has begun to lose its academic flavour. As the potential of computer-mediated communication has been realised and better, easier-to-use software has been derived, the Internet has attracted commercial "Internet Providers". These providers allow access for people who otherwise would not be able to use the Internet.

Who Runs it?

The Internet has little formal organisational structure. Due to the need for control over domain names and node numbers their allocation is carefully controlled by the Internet Network Information Centre (InterNic), based in the US, with some delegation, as described in Chapter 2.

Technical decisions are made by consensus. Ideas may be suggested by anyone; they are discussed and formalised, and if it is agreed that they are "good", may be adopted by the Internet Architecture Board, a self-perpetuating group of volunteers. This process of "floating" (proposing in an informal and deniable way) and discussing ideas is carefully controlled. This practice of floating an idea and seeking a consensus is the reason that the Internet standards documents are called "Requests for Comments". In the past some implementations of Internet software have failed to observe all the fine detail of RFCs, so that they have been honoured more in the breach than in the observance; modern implementations are much more particular about technicalities.

However, you may still find odd problems caused by, for example, mailers disagreeing about the format of a header line.

Sometimes this lack of formal organisation makes the Internet difficult to manage. It is likely that in the future a greater degree of central control may be established, if for no other reason than to get Government money invested in increased communications bandwidth. It seems likely that with the current boom in Internet use, the policy of the U.S. Government to institute a data "superhighway", and increased use by private rather than academic sites, that the organisation of at least the "backbone" of the Internet will become more formal, and more carefully monitored so that "the user pays".

How Big is it?

In truth, nobody knows how big the Internet is. All sorts of figures are tossed around, but any two estimates of the size may easily differ by a factor of five. One estimate is that in the second half of 1994, about 100,000 new users were being added per month! Due to the distributed way in which the Internet is managed there is no simple answer to this very simple question. It is obvious that the growth rate, whether measured in terms of the number of connected sites, number of users or traffic volume, is very high.

In practical terms, users can assume that all universities in the USA, Australia, Canada, NZ, UK, Asia and Europe, can be reached through the Internet. So can most governmental research organisations, tertiary institutions, many commercial database suppliers, private research institutions, libraries, governmental agencies and some hobbyists bulletin boards. There are sprinklings of secondary-level educational institutions, corporations, and an increasing number of public internet suppliers. There are also some privately owned systems. There is less internet penetration in Africa and South America, but the growth rate is high.

What is Available?

The major "services" provided via the Internet are:

- Person to person contact, through electronic mail.
- Discussions between groups of people, through Usenet News, mailing lists and on-line conferencing.
- Access to databases, in which the Internet is used as the medium to connect to commercial database providers.
- Access to library catalogues (Online Public Access Catalogue or "OPACs").
- The ability to transfer files privately between computers.
- The ability to uplift files from publicly available archives of software and data.
- The ability to execute programs on remote computers.

Not every Internet node provides all the available services; the most common are electronic mail and Usenet news. Because of these limitations there are a number of *ad hoc* arrangements which allow you to use Email for most purposes when all else fails.

Where is the Science?

The Internet is widely seen as a great resource for technofreaks and computer geeks, and increasingly as a playground, but what's in it for the busy scientist? At this point you should quickly glance at the table of contents for the second part of the book, and see the variety of sites, services and opportunities. The "information superhighway" has been noticeably hyped, and over-sold to the general public; to the scientist it is a real tool.

Other Networks

There are several other forms of electronic communication network which aren't part of the Internet. Hobbyist bulletin boards frequently exchange material by belonging to FidoNet; CompuServe, CIX and BIX were initially stand-alone information services. These services have partially merged with the Internet, and can be reached through the Internet, or at least exchange mail. There are mail networks, typically run by telephone companies, which are closed, in the sense that subscribers can only mail each other. Typical examples are Starnet, and Telecom Gold. These are not generally related to the Internet.

Further Reading

Printed

There are many general books on Internet topics. Many become out of date very quickly, and few devote much space to topics of scientific interest. The following may be useful:

Browne, S. 1994. The Internet via Mosaic and World Wide Web. Ziff-Davis Press, Emeryville, CA.
Engst, A. 1994. Internet Starter Kit for Macintosh. Hayden Books, Indianapolis, IN.
Gaffin, A. 1994. Everybodys Guide to the Internet. MIT Press, Cambridge, MA.
Krol, E. 1994. The Whole Internet User's Guide & Catalog 2nd Ed. M. Loukides (Ed.). O'Reilly & Associates, Sebastopol, CA.
LaQuey, T. and Ryder, J.C. 1993. The Internet Companion: A Beginner's Guide to Global Networking. Addison-Wesley, Reading, MA.
Raymond, E.S. 1991. The New Hacker's Dictionary. MIT Press, Cambridge, Mass.
Smith, R. 1994. Navigating the Internet. Sams Pub.

On-Line Documents

Documents available on-line may be more up to date than printed material. The locations given are not the only places these files may be found. Where possible, the home location is given.

de Presno, O. The Online World (A shareware book)
`ftp://ftp.eunet.no/pub/text/online.txt`

Hardy, H.E. 1993. The History of the Net.
`umcc.umich.edu/pub/seraphim/doc/nethis8.txt`

NCSA Education Group, The Incomplete Guide to the Internet,
available as a Microsoft Word document or as a Postscript file.
`ftp://ftp.ncsa.uiuc.edu/Education/Education_Resources/`
`Incomplete_Guide/Dec_1993_Edition/MS_Word/*.hqx`
(the document is in several files, for Macintosh)

`ftp://ftp.ncsa.uiuc.edu/Education/Education_Resources/`
`Incomplete_Guide/Dec_1993_Edition/postscript/*.ps.Z`
(several compressed Postscript files)

Yanoff, S. Speial Internet Connections
`ftp://ftp.csd.uwm.edu/pub/inet.services.txt`

Yanoff, S. Inter-Network Mail Guide. (How to mail that odd address from your odd address)
`http://alpha.acast.nova.edu/cgi-bin/inmgq.pl`
`ftp://ftp.csd.uwm.edu/pub/internetwork-mail-guide`

There are literally thousands of other sources; each of these lists many others.

From: *Internet for the Molecular Biologist.*
ISBN 1-898486-02-6 ©1996 Horizon Scientific Press, Wymondham, U.K.

2

MACHINES, USERS
AND ADDRESSES

Gordon Findlay

Introduction

The purpose of this chapter is to introduce the principle of machine addressing. The chapter describes how computers are addressed using a numbering system, how that numbering system is made easier for us to understand using a hierarchical naming system and how several users on one machine can be addressed.

A Net of Networks

Computers communicate with each other by passing packets of data around. Precisely how the packets of data get from one machine to another is unimportant; all that really matters is that the packets are in an agreed format, so that each computer can interpret packets in the same way. Whether the packets are passed around by hard-wired connections, over ethernet, through modems, fibre-optic cables, satellites or under-sea cables is, from the point of view of a user, irrelevant. There may be a noticeable difference in speed, but that's probably all.

   The Internet is an Inter-network; that is to say it is formed by connecting together other networks. A typical situation is a local-area network within a building, which is connected at one point to another network somewhere else, which is itself connected to other networks, and so on. Provided that there is a common protocol for the assembly of data packets ("datagrams" in the jargon), computers on these networks can effectively communicate as if they were all directly connected.

   Packets of data need not travel directly from one machine to another. They may be passed through several machines on the way. As a test I've just checked to see the route that a mail message from this machine in Christchurch, New Zealand takes to a friend's machine in Montreal, Canada. Each datagram (packet of data) making up the mail message passed through 17 other machines on the way!

   Obviously there is the potential in a situation like that for one of the intermediate machines to be "down" or not functioning. Routing packets through the Internet is a very complex topic; suffice to say that the network protocols are designed to find another route if a machine is down. However, occasionally you will find that a particular machine is inaccessible. These "outages" are usually brief. The capacity of the links (or "bandwidth") varies. The Internet backbone machines (the ones which handle most traffic) naturally have the fastest links.

Numbers

Every machine on the Internet has a number. This number is the machine's identifying address and must be unique. These numbers are written in a "dotted decimal" form; this paragraph has been typed on machine number 192.101.16.7. If any other machine anywhere is also using that number, there will soon be terrible confusion. The need to make these numbers unique means that they are issued from a central point, the InterNic, or from a delegated block of numbers controlled by a secondary authority, such as a country NIC (network information centre). Rather than have the NIC record details of every machine, which would be impossible ("I'm just moving my laptop into the next room: phone the NIC would you?") the NIC records details of networks, and issues a block of numbers to the Network Manager of an institution. In this example, the Network Manager has been given the block of numbers 192.101.16.0 to 192.101.16.255, and can allocate and change them as he wishes without reference to anyone outside the organisation. The Network Manager is said to have authority over those numbers delegated to him.

Names

As no Network Manager can keep track of addresses in dotted decimal form, never mind expecting users to be able to, it was very soon realised that easier ways than numbers would be needed to identify machines. Every machine could have been given its own single-word name; and indeed BITNET worked that way. This required someone to keep track of all of the names with the result that it was not unusual for the changes in one day to be too many to process before the next day's changes started. Enter the "domain name".

Let's take an example of a computer's name: BEAST.CHMEDS.AC.NZ

Read it backwards: it is a machine in the .NZ domain, which covers the whole of New Zealand. Within New Zealand, it is a machine belonging to a site in the .AC subdomain - the academic community. The particular site in the academic community is CHMEDS - Christchurch Medical School. And within the group of machines in that noble institution, is one called BEAST. Naming machines in this "hierarchical" manner means that control can be decentralised, and devolved downwards. Sticking with the example, control over .NZ is delegated to the networking group at the University of Waikato. Subdomains of .NZ are delegated to different organisations to spread the load, so that .GOVT.NZ (New Zealand governmental or quasi-governmental organisations) is controlled by Victoria University, Wellington.

Just as a Network Manager has control of a block of numbers, he has control over names at the bottom level: the machine called BEAST can be renamed BEAUTY.CHMEDS.AC.NZ without reference to anyone outside of CHMEDS; liaison is only needed when changes occur at higher levels. To complete the picture: when a user is to be addressed at a machine simply add the username and an @ symbol to get the user's Email address, such as;

`TOOTH.FAIRY@BEAST.CHMEDS.AC.NZ`
(please note that this person is not reachable by Email - yet).

Major Domains

Most countries are a top-level domain;

* `.nz`   New Zealand
* `.uk`   United Kingdom
* `.fr`   France
* `.us`   USA
* `.de`   Germany
* `.es`   Spain
* `.au`   Australia
* `.ch`   Switzerland
* `.za`   South Africa

Most countries have subdomains such as;

* `.ac`   Academic
* `.gov`   Government
* `.co`   Company
* `.com`   Commercial organisations.
* `.org`   Other organisations.
* `.gen`   General, eg bulletin boards.

There are also a number of top level domains which are not country-based. These domains are used almost exclusively by USA sites;

* `.edu`   Educational, eg `JHU.EDU` for John Hopkins University.
* `.mil`   Military, eg `ARMY.MIL`.
* `.gov`   Governmental, eg `NASA.GOV`.
* `.com`   Commercial.

These domains predate the invention of the .us domain and are probably useful as such a large number of sites are in the USA, although controversy about who should be able to belong to them breaks out periodically.

What's My Number?

The invention of the domain name system makes it easier for people to use the network, but the computers still have to work with numerical addresses. The translation from one to another is done using a very clever piece of distributed intelligence, called the

Domain Name System (DNS). Let's take an example. Suppose that a computer has to contact another, and has only the name CONSULTANT.MICRO.UMN.EDU with which to work. It sends a request to a piece of software on a designated machine (the "name server") which knows about the domain MICRO.UMN.EDU, and asks for the address of CONSULTANT in numerical form. If, as is likely, the requesting machine doesn't know which machine is the name server, the request can be passed to the name server for the next higher level domain (UNM.EDU) or if that is also unknown to the name server for the top-level domain (.EDU). As a last resort, a small number of "root" name servers are universally available, and can pass a request down to the next appropriate place.

Each higher-level name server can give the address of the next server (looking downward) to query. The hierarchical nature of the domain system means that each nameserver only needs to know about a very few others, and the maintenance of the databases is spread over a large number of sites, making it more manageable. Any changes of name or number at the local level need to be reflected in the local name service database; the software is designed so that these changes will gradually spread outwards, on a "need to know" basis. There are many complications carelessly brushed aside here - leave the details to your Network Manager.

Further Reading

Printed

There are many general books on Internet topics. Many become out of date very quickly, and few devote much space to topics of scientific interest. The following may be useful:

Browne, S. 1994. The Internet via Mosaic and World Wide Web. Ziff-Davis Press, Emeryville, CA.
Engst, A. 1994. Internet Starter Kit for Macintosh. Hayden Books, Indianapolis, IN.
Gaffin, A. 1994. Everybodys Guide to the Internet. MIT Press, Cambridge, MA.
Krol, E. 1994. The Whole Internet User's Guide & Catalog 2nd Ed. M. Loukides (Ed.). O'Reilly & Associates, Sebastopol, CA.
LaQuey, T. and Ryder, J.C. 1993. The Internet Companion: A Beginner's Guide to Global Networking. Addison-Wesley, Reading, MA.
Raymond, E.S. 1991. The New Hacker's Dictionary. MIT Press, Cambridge, Mass.
Smith, R. 1994. Navigating the Internet. Sams Pub.

On-Line Documents

Documents available on-line may be more up to date than printed material. The locations given are not the only places these files may be found. Where possible, the home location is given.

de Presno, O. The Online World (A shareware book)
`ftp://ftp.eunet.no/pub/text/online.txt`

Hardy, H.E. 1993. The History of the Net.
`umcc.umich.edu/pub/seraphim/doc/nethis8.txt`

NCSA Education Group, The Incomplete Guide to the Internet,
available as a Microsoft Word document or as a Postscript file.
`ftp://ftp.ncsa.uiuc.edu/Education/Education_Resources/`
`Incomplete_Guide/Dec_1993_Edition/MS_Word/*.hqx`
(the document is in several files, for Macintosh)

`ftp://ftp.ncsa.uiuc.edu/Education/Education_Resources/`
`Incomplete_Guide/Dec_1993_Edition/postscript/*.ps.Z`
(several compressed Postscript files)

Yanoff, S. Speial Internet Connections
`ftp://ftp.csd.uwm.edu/pub/inet.services.txt`

Yanoff, S. Inter-Network Mail Guide. (How to mail that odd address from your odd
address)
`http://alpha.acast.nova.edu/cgi-bin/inmgq.pl`
`ftp://ftp.csd.uwm.edu/pub/internetwork-mail-guide`

There are literally thousands of other sources; each of these lists many others.

From: *Internet for the Molecular Biologist.*
ISBN 1-898486-02-6 ©1996 Horizon Scientific Press, Wymondham, U.K.

3

INTERNET SERVICES

Gordon Findlay

Introduction

This chapter sets out to describe the sorts of ways that information is distributed across the Internet. There are many different forms of communication and information sharing available, each with their own strengths and weaknesses. Traditionally, each type of service has been handled by a separate piece of software (for example, gopher, FTP...). Increasingly the more sophisticated packages, particularly WWW browsers such as Mosaic and Netscape, provide a more integrated environment, with many services available through the one piece of software.

The traditional text-based interfaces are being replaced with graphical interfaces, in either the Macintosh or MS-Windows environments, or running over X-Windows. However, the text-based services still have some advantages, and will be with us for some time yet, for three reasons. Some Internet users have only text-based terminals available. Text-based systems have a lower bandwidth requirement, and when users must pay by the volume of data moved, text interfaces will be much cheaper. Over slow links, text interfaces, with the much lower amount of data to be transmitted, are much more efficient. Of course, text based systems require more effort to master, and cannot give access to information in graphical form. The software used to implement these services differs markedly from machine to machine. For example, a Macintosh program will, if it is any good, make full use of the special features of the Macintosh user interface. Obviously the same program would be useless for a dumb terminal connected to a Unix computer.

Electronic Mail

The genesis of the Internet was the desire to pass mail messages electronically. Email is undoubtedly the most frequently used of the Internet services (although it probably isn't the biggest in terms of the volume transferred). Email is the most ubiquitous of the Internet services; for this reason it is often possible to perform other functions such as file transfers by Email.

The Email Message

Suppose you want to send me a message quickly and cheaply. It can be as simple as typing this:

```
mail gordon@chmeds.ac.nz
Subject: Lunch tomorrow
I want to buy you lunch tomorrow.  Can I
meet you at 1pm?

Cheers.
```

The message (to which the answer is always "Yes") makes its way to my computer; if I am using my computer (logged on) I'll be told that a new mail message has arrived. If I'm not logged on, I'll be told that there is a message for me when next I do log on.

Mailers

There are very, very many pieces of software around to use for sending mail. These mailers differ in detail. Some provide forms on screen for you to type in the name of the sender and recipient, with sophisticated editors to compose the message. Some know about the full range of Internet addresses; others need a little help. A few mailers are able to tell you whether a message you sent someone has been read yet. This notification is usually only possible on a local system, not across the Internet. In general, Email is just like paper mail: you send it, but unless you get a response you can't be sure it reached its destination. Email is pretty reliable though: lost messages are very uncommon.

Addresses

It is difficult to know where to begin with the problems encountered with addresses. The most common address, as described in Chapter 2, is the usual Internet form of

```
user@machine.site.domain
```

Local addresses (for people using the same system) are usually just a user name. It's difficult to know where to find people's addresses. There is no Email analogue of the telephone directory, although organisations may keep local directories. The best way to get someone's Email address is to ask directly. There are some directory services (X.400, Netfind) but these are of limited use. Subject associations frequently publish lists (of varying degrees of completeness). The Usenet group bionet.addresses may be useful.

Most sites have a special address of postmaster. This is the poor guy who gets all the problems to deal with. Postmasters may be prepared to reveal a user's Email address in an emergency, but don't use it as a routine means of making initial contact. Some postmasters get several hundred messages per day and are understandably loath to accept any additional hassle.

BITNET Addresses

Many American sites still give addresses in the old BITNET style, even when they have a perfectly acceptable domain-style address. A BITNET site name is just a machine name rather than a hierarchical, domain style name. Distinguishing a BITNET address is easy: if the part after the "@" does not contain any full-stops (periods) then the address is in BITNET. Many mailing list addresses are in this form, for example

`listserv@umupvh1`

Some mail software is smart enough to know what to do if you simply add .BITNET as the domain like this: `listserv@umupvh1.bitnet`. Other software can do the right thing if you;

> (a) change the "@" to "%"

and

> (b) add an Internet-Bitnet gateway machine:
> > `listserv@umupvh1` becomes
> > `listserv%umupvh1@cunyvm.cuny.edu`
> (`cunyvm.cuny.edu` is one gateway but there are others).

Signatures

You'll soon find that most people have special "signature blocks" at the end of their messages. A signature block is a short text file containing your name, electronic address, street address and other information, and frequently a short witticism, quotation or personal statement. Most mailers can automatically add your signature to the end of every message you send. Two traps: it is a sure sign of a "newbie" to make the mistake of having your signature included twice. A "doubled sig" really does look quite naive. Keep signature blocks short: four lines is often recommended as a maximum. There is no need for fancy pictures or imitations of your signature; what looks cute or clever the first time soon pales, and a long signature block is a waste of bandwidth.

Mailing Lists

One of the most common ways to circulate information and promote discussion electronically is through the use of a mailing list. As Email access is often easier than other forms of electronic communication, mailing lists are becoming increasingly common. There are now many thousands of mailing lists, on all sorts of topics, from Library Administration to Audi Quattro cars!

The functioning of a mailing list is simplicity itself. Someone who has something to say, or a question to ask, or an answer to someone else's question simply sends an Email message to a particular address (the list address) in the usual way. A computer program makes copies of the message, and sends one to each list subscriber. These messages turn up in each subscriber's personal mailbox, just like private Email.

For example, a mail message to `CATFISH@CHMEDS.AC.NZ` will be circulated to all members of the (fictitious) CATFISH list.

There are many varieties of mailing list. Some lists are moderated. Mail sent to a moderated list is reviewed by a particular person, or "Moderator", before it is forwarded to the members of that list. In this way repetitive questions, erroneous messages and junk mail are eliminated. Lists may appear in digest form: contributions to discussions are grouped together and sent out in batches. Some lists are very active: NOVELL-L averages about 90 messages per day. Others are very quiet, with only a few messages per week.

There are Two Addresses

Of course, there is some administration required to make a mailing list work. There are mechanisms for joining a mailing list ("subscribing"), leaving it ("unsubscribing", a horrible neologism), seeing who else is a member of the list ("review") and perhaps for obtaining copies of previous messages from an archive somewhere. These sorts of tasks are handled of course by Email: to an administrative address. In many, many cases the administrative address is `LISTSERV@somewhere`. In other cases the administrative address is formed by adding `-request` to the mailing list name. In any case, sending administrative requests to the list itself will serve no purpose (other than making all the list members wonder what is wrong). The distinction between the address of the list itself, and the administrative address for the list cannot be emphasised enough. In the example above, `CATFISH@CHMEDS.AC.NZ` is the list address; the administrative address is `LISTSERV@CHMEDS.AC.NZ`.

Joining a List

To join a list you need to know the administrative address: commonly `listserv` but not always. If you don't know the address, stop and find out: don't use the list address. Once you have the address, send a message consisting of just the line:

`SUBSCRIBE` listname your-first-name your-surname

to the administrative address.

For example, if your name is Wendy Bloggs, and you wish to join the FROGS-L list, send the message

`SUBSCRIBE FROGS-L Wendy Bloggs`

to

`FROGS-L-REQUEST@CHMEDS.AC.NZ`

This line belongs in the body of the message, and you may as well leave the subject blank, as the computer programs which handle the subscribing will ignore the subject anyway. The first message you get back will be the most important! It will be from the list's "owner" or manager, and it will give you instructions for how to leave the list, how to tell the list that you are on holiday and don't want mail, how to send messages, how to get the list of subscribers, and whether there is an archive of previous messages which you can gain access to. Keep this initial message! Every day literally dozens of messages are sent all over the world by people who have forgotten how to sign off a list or perform some other routine task. Some high-volume lists require an acknowledgement of this initial message to confirm that the list software found your correct address.

Leaving a List

Normally you leave a list by sending a similar message to the administrative address for the list:

UNSUBSCRIBE listname

It is important that you do unsubscribe from lists if your Email address is about to change. Failure to do so can lead to "mailer loops", in which an error message goes back to the list, and is copied to your (now obsolete) address, which generates an error message ... That should not happen, but often does, as setting up a mailing list to properly handle the wide variety of mailers around the world is virtually impossible, especially as many sites have imperfectly configured mailers.

Sending a Message to the List

Simple! Just compose the message, and send it as a normal message to the list address. It will be duplicated and sent on over the next few hours depending on the size of the list. Many mailing lists are clever enough not to waste time and bandwidth by sending a copy of your own message back to you. It is often possible to override this behaviour if you wish: read the initial message again.

List Archives

Many lists have previous messages archived somewhere for retrieval, by Email, FTP, Gopher or other means. Some have properly maintained indexes while others have none. Accessing the archives will usually be described in the initial message that you receive. If not, send the one-word message HELP to the administrative address.

List Administrivia

Attention to a few administrative details will greatly enhance your use of mailing lists. Remember that the lists are driven by sometimes remarkably bone-headed software, which may not always do the right thing.

Holidays

If you are going on holiday, and don't wish to be greeted by hundreds of messages when you return, you can temporarily suspend mail from most lists by sending the message SET listname NOMAIL
Reverse this by sending SET listname MAIL when you want to start getting mail again.

Vacation mailers

If you have a vacation program running, which notifies correspondents that their message will be ignored until you return. Be sure to SET NOMAIL for all your mailing lists, other list members don't want to see your vacation message even once, never mind once for every list message over a period.

Who's on the list

Generally the command REVIEW listname returns a list of all the subscribers to the list. (It is possible to conceal your subscription from others if you wish).

Etiquette

There are a few traps to watch for once you are receiving messages. Be sure you don't reply to the whole list when you mean to write just to the author of a message (be careful of the mailer command REPLY). Usually it helps the flow of a discussion to include relevant parts of a message that you are commenting on; it is considered bad form to include a whole message and just add "I agree". Replying to administrivia sent by accident to the list should be left to the list owner or moderator; otherwise the list will soon become clogged with junk mail. Remember that as in all computer-mediated communication, it is difficult to express irony in Email. There is no body language, so ironical or comic comments can simply appear abusive. To avoid this problem authors may use "smileys". These are small pseudo-graphics that try to relate the intent behind the message (see the glossary).

Directories of Mailing Lists

There are many hundreds of lists; an almost complete directory of them is given in the file MAILING.LISTS, available for FTP from bliss.chmeds.ac.nz in the Internet directory. Partial directories are posted to Usenet News in the news.answers group monthly.

Mail Servers

Because Email is so ubiquitous, and is available at even very badly connected sites, many program archives and databases are made available for searching by using a mail server. An Email message sent to one of these servers will automatically return (some time later) the requested files or indexes as Email. This is convenient, but does have limitations. For example, large files must be split into mail-sized bits and most files must be encoded to make them mailable. The number of things that can go wrong is immense. These servers vary considerably in operation, and it isn't possible to give any general rules for their use. Send the HELP command first.

As sequence databases are very commonly searched by Email, a number of special programs have been developed as "front ends" for the use of mail servers such as EMBL, GenBank, PIR etc. These simply take users' requests in an interactive session, format and send an appropriate (and correct!) message to the mail server, and perhaps display the retrieved data in a convenient form.

Electronic Journals

Although in its infancy at present, Email distribution of journals is becoming important. Typically these journals are low-volume, unrefereed publications produced to cover a niche that is small, rapidly developing and appeals to people with a bias in favour of using technology. Subscribing to an electronic journal is much like joining a mailing list. The difficulty is in finding them to subscribe to. It is most likely that you'll come across them by such means as mentions in News, on Email, through Gopher or by word of mouth.

Usenet News

One of the most controversial, most misunderstood, most useful yet most infuriating of the information sources available to the Internet user is Usenet News. Usenet distributes a vast amount of information every day; along with, it must be said, a vast amount of garbage and misinformation. Email is undoubtedly the most easily available of the Internet services, with News in second place. News has four features that make it so popular:

- It requires little in the way of equipment to receive and read news: any dumb terminal or PC in text mode will do.
- It is a lot of fun, especially to technically oriented people.
- It has a very wide coverage of topics, from the information-theoretic aspects of biology to the Society for Creative Anachronism.
- It is fairly easy to get a new newsgroup started to discuss absolutely any subject at all. In fact, whatever the subject is, there is likely to be already a newsgroup covering it.

Anything, and everything, is discussed somewhere in Usenet ("Usenet", "News" and "Usenet News" are used almost interchangeably). It's a wonderful place to get lost in, and to find gems in and is frequently used as a source of pointers to other information resources.

What is it?

It's difficult to describe! News articles are "posted" by individuals at any of thousands of sites, and are propagated to other sites in batches over a few hours. Perhaps twenty-four hours suffices to cover most of the globe. These news articles are tagged with a subject, and placed in groups, so articles relating to vegetarian cookery are found in the newsgroup rec.food.vegetarian (the full-stops are part of the group name). Anyone reading the news who wanted to read only about vegetarian cookery would read only the articles in that group.

A Multitude of Newsgroups!

There are well over 4000 newsgroups commonly distributed world wide, and over 8000 with restricted distribution. The aggregate traffic is over 1500 MB in 800,000 articles per month. It is not possible to read them all! The newsgroups are divided into hierarchies, in a tree structure (like disk directories). There are seven major hierarchies:

- comp: Topics about computers, software, operating systems, programming and computer science.
- misc: Topics which don't fit other hierarchies, such as investments, job-hunting and personals.
- news: Groups concerned with the operation of the News network, and software for reading and propagating news.
- rec: News groups relating to recreational topics.
- sci: Serious level, and increasingly now, popular level, scientific topics.
- soc: Groups for social issues, and socializing, including groups about many world cultures.
- talk: Groups for debating all sorts of topics (usually if not always without significant progress towards resolution).

The above groups are those circulated most widely. In addition, there are other hierarchies which are less widespread:

- alt: The alternate groups: ranging from "Married with Children" to sex. Many of these groups (e.g. the alt.folklore groups) contain serious content; others (e.g. alt.flame and alt.swedish.cook.bork.bork.bork) do not; they are "for amusement purposes only".
- bionet: Groups for biological and biomedical topics. These are the most useful for molecular biologists. Not all sites carry the bionet groups; if your site does not (and you cannot persuade the management that they ought to be) then most of these groups are available by mailing list.

- bit: These "bitnet" groups are usually a redistribution of the contents of a mailing list: for example bit.listserv.novell-1 is a redistribution of the NOVELL-L mailing list in newsgroup format for those who prefer it this way.
- biz: Business oriented groups.
- vmsnet: Topics relating to OpenVMS, an operating system for Digital Equipment's VAX and AXP machines.

There are many other distributions, such as aus groups, which are intended for Australian users, and even distributions down to the city or university level. These geographic distributions frequently "leak" outside their proper domain.

Restricted News Feeds

Not all groups are available everywhere. There are several reasons why a site may not carry some groups - lack of space to store them is a common one. Other reasons include lack of bandwidth to handle the traffic, and institutional policy (for example, EEO policies frequently cut access to groups carrying erotica).

Newsgroups and Mailing lists

Several of the larger mailing lists are gatewayed into newsgroups. This means means that anything posted to the mailing list "automagically" appears in the newsgroup as an article, and anything posted in the newsgroup is distributed around the mailing list. If you have a choice you will probably find that it's easier to read a thread in a newsgroup than to read disjointed mail messages.

Some Important Newsgroups

There are three important groups for new users:

- news.answers
- news.newusers.questions
- news.announce.newusers

Read these groups for a few weeks to become familiar with the way the system operates, both at a formal level and informally, and the range of material available. In the bionet hierarchy, bionet.announce contains important announcements relating to these newsgroups. If your institution has a local newsgroup, be sure to keep up with it.

Posting News

Having your own say is the fun part of Usenet. Firstly, distinguish between a posting, which is a new article, and a follow up, which is an article commenting on or answering questions in another. If you are asking a question, be crystal clear in what you ask, and provide all relevant information. Don't just say "I got an error message" - give the exact message! Ambiguous or vague questions will, if you are lucky, attract no replies. If you are unlucky, they will attract flames (verbal attacks) and abuse. Mind you, some groups are just for flaming, and the more inflammatory you are in those groups the better. Badly written and punctuated articles with obvious grammatical errors won't attract a useful response. Don't write in ALL CAPITALS; it reads as if you were SHOUTING.

It is important that you pick a useful subject line, otherwise your article will be lost in the mass and not read. If it's a follow up, use the original article's subject, prefixed with Re:. Some topics have become known as "religious wars". "The Foobar is a better computer than the Bazqux", Windows versus Macintosh, my editor versus your editor, Biblical statements that the Earth is flat, the evidence against evolution and similar topics simply use up bandwidth and are quite incapable of producing rational argument or information in this medium. Often posters ask for replies to be sent by Email. If you do this, it is courteous to post a summary of the replies to the newsgroup after a decent interval, so that other readers who are interested in the subject can benefit as well.

Remote Computing

Another important early function of the Internet (or of the networks that became the Internet) was the use of the net to connect users to computers at another site, and use them as if physically there. The initial impetus was to give researchers who were widely dispersed physically the ability to use the few supercomputers that were available. This is still done, but less commonly now. A major application of this sort of remote computing is to access library catalogues at other sites; another is to search literature databases such as Dialog. The main protocol used for this sort of remote computing is Telnet. Connecting to another computer is as easy as typing

```
TELNET computer.site.domain
```

(where computer.site.domain would of course be the Internet name of the computer you wanted to log in to).

Assuming that the remote site is reachable (no downed systems between you and it), you will soon get the login sequence for the remote machine, just as if you were there. Obviously you need an account name and probably a password in order to complete the login. Thereafter, you are talking with the remote machine, and all its procedures must be followed. Log off in the correct manner to break the Telnet connection. Naturally, the further away your system is from the remote system, the slower the response will be. Depending on the speed of your link, and the amount of traffic over the links between you and the machine, you may notice anything from a small deterioration in speed to unacceptable delays.

Telnet has a few traps for the unwary user.

- It can be expensive. Many small packets must be sent rather than a few larger packets. Each packet has an overhead associated with it, and these overheads can soon mount up.
- Make sure you know how to log off the remote system before you log on to it. Most library catalogues display instructions about this only at the start of a session, so make a note of it then.
- CaSe MaTtErS. Many computer systems are case sensitive which means that if the command is "ls", and you type "LS", it won't be recognised.
- Terminal types. There are many different types of terminal, and the match between the terminal (or program) you are using, and what the other system expects, may not be perfect. If you are using a PC, the terminal emulation provided by the software may not be an exact match for the real terminal being emulated.
- One serious problem which frequently occurs is that different terminal types require different characters for the backspace operation. A little experimentation may be needed to resolve this - try the backspace, Delete, left arrow and control-H keys to begin with.

Moving Files

It is possible to Email files, but most file transfers are done using FTP. If FTP access is impossible, it may be possible to simulate it using a mail or file server.

Moving Files with FTP

File Transfer Protocol (FTP) is a program, and a protocol, for moving files between computer systems across the Internet. It is convenient, readily available, virtually error-free, and can be a most unfriendly thing to use. It must be frankly admitted at the beginning that FTP can involve all sorts of technical troubles. Despite this, FTP is worth persevering with, as it is frequently used to obtain resource guides and information about other tools and resources. With luck you will be using a graphical package to FTP with. If not, you will find that a small collection of Unix commands are required. However you get the files to your system, getting them there is only the beginning.

When to Use FTP

Use FTP when you need to transfer files between machines - either to your machine from a remote site or from your machine to another. This may be by transferring to or from a normal account on the remote machine (in which case you need a user name and password, just as if you were logging on) or to a special, "anonymous" account which is open to all. The latter, anonymous FTP, is often used as a method of making documents, programs and data available to anyone who can pick them up.

File Formats

There are two types of file, as far as file transfer is concerned: text files and binary files. A file transferred in the wrong "mode" will be unusable and it can be difficult to see which mode is appropriate. ASCII mode is used for text files only. Often the file name will be "something.TXT", or ".LST", or ".ASC". The README file will always be ASCII. In general, files which are intended to be read "as-is" by a human, rather than manipulated further by a computer, are ASCII. Postscript files, usually with names ending ".ps" are also transferred in text mode. Some data files are text, others are binary. Wordprocessor documents are generally not ASCII: use binary mode with them. All program ("executable") files (.exe), compressed files (.Z, .z, .gz, .zip, .arc, .hqx etc), graphic files (.tif, .gif, .pic, .pcx, ...) and many data files must be transferred in binary mode. If there is a README (or readme or read.me or ...) file it is usually worth grabbing that before trying to work out what files to get and the mode to get them in.

Compression

To save disk space, which is a scarce and expensive resource in many places, and the cost of transmission, files are often compressed by one means or another. Often, a collection of related files will be compressed into one "archive" file to make handling easier. There are many programs used for compression, and new ones are being developed all the time. Compressed files are identified by their name ending in .z, .Z, .zip, .pak, .arc, .lzh, .lha, .arj. .boo, .zoo, .gz, .gzip or one of many other extensions, which may reveal to the experienced eye the type of compression used. It is futile to describe how to decompress a file here. The software available, and its use, vary considerably between machines. Unless you are already familiar with such arcana as "zcat foo.tar.Z | tar -xvf" you will need to get help from someone who knows the software and operating system which you are using.

Anonymous Archives

Several sites have set out to collect material relating to a particular interest or field. For example, the site

`WUARCHIVE.WUSTL.EDU`

usually known as WUARCHIVE has a very large collection of public domain and shareware software. These collections are often made available for anonymous FTP as a service to the Internet community. These archives may, like WUARCHIVE, be very large and general purpose; or may be small and specialised; for example the collections of Internet and Paradox-related material to be found on `BLISS.CHMEDS.AC.NZ`. To reduce the impact on the sites concerned of this generosity, and to conserve communication bandwidth, mirror sites duplicate a collection. It is considered bad form to use a popular site if a closer mirror site can be accessed. Anonymous FTP sites are accessed by using the user name "anonymous", and your Email address as the password. (So your use of the site isn't really anonymous). This allows access to anyone, without having to have an account on the host machine.

Gopher

Searching, finding and retrieving information from the vast range accessible via the Internet has become increasingly difficult, particularly as the Internet has become less the domain of computer people and more the domain of information users. The Gopher initiative was developed to address several points;

- The problems that most users had with FTP.
- The desire to make more types of information (such as images, audio and video) available.
- The need to make it possible to browse and find information without knowing exactly where it is.

The ideas behind Gopher originated at the University of Minnesota, and the team there still coordinate much of the development work for Gopher software, although the development work is being carried out world-wide. It is not universally available, for example, it is unlikely to be available if you access Internet through a bulletin board. If it is available, it is one of the more powerful tools in your information arsenal. Gopher is comprised of two parts. The software which you run to search and retrieve is a Gopher client and the software which answers your requests and serves up the data is a Gopher server. We are only interested in Gopher clients here.

Starting Gopher

To start a Gopher session, just type Gopher, or click on the Gopher icon. You will be presented with a menu, which you can navigate with arrow keys, or (if you are using a rodent-aware Gopher) a mouse.

This menu presents a variety of types of information, including:

- Text files.
- Directories.
- Collections of documents.
- Search items: a special type of directory which is searched by keyword. Only the documents which match the search criteria will appear subsequently in the directory.
- Telnet sessions: an item which will connect you to another system such as a remote library catalogue.
- Phone books: a special type of search, which allows the user to look for names, addresses, and especially Email addresses.
- Multimedia items: useful if you have a viewer, although many of us don't yet.
- Links to other Gophers. Gophers can retrieve documents from other computers for you, or connect you to other Gophers with interesting resources.

It is this last characteristic which gives Gopher its name: Gopher "burrows" through the Internet making connections as it needs them, just as the eponymous animal burrows underground. All manner of information is available on Gopher servers. Weather forecasts, library catalogues, databases of all sorts, sequence files to download or search, recipes, newspapers, Usenet News, electronic journals and free software to mention a few.

Which Gopher?

To some extent it isn't necessary to know just where Gopher is connecting to first, as each Gopher server knows about (most of) the others. Very often the initial connection is made to the "mother Gopher" at the University of Minnesota. From there you may connect to any other "Gopher-hole" that you wish. Your Gopher may be set up to initially connect to some other important site. Many Gophers which access biological information connect initially to a Gopher server at the Welch lab, John Hopkins University. If you know that you want to use a Gopher archive at a particular place you may connect directly, for example, to GOPHER NS.RIPE.NET. This is an important gateway to European sites, and has pointers to many important data depositories.

Navigating in Gopherspace

It is very easy to become lost in Gopherspace. This is the major limitation on usefulness of Gopher: it is very like wandering around a library in which all the books are neatly shelved, but in random order. The usual way of recording a gopher location is to give the site first connected to, then the menu choices, in order. As an example, there is a catalogue of software for agricultural scientists at the location:

```
gopher merlot.welch.jhu.edu

13 Search and retrieve software
  7 Search and retrieve agricultural software
   1 Search for agricultural software (NCSU)
    2 California
```

Each line records the next menu item to choose in turn (Gopher menu lines are always numbered). The similarity of these menu items is an indication that a jump to another machine has occurred during the search. Unfortunately not available in all Gophers, bookmarks provide a very useful way to keep track of that very interesting location with just the information you need. In effect, they provide a way for you to select your own dinner from the smorgasbord of items on Gophers all over the world.

Bookmarks are stored in a special file called a resource file, with a name something like .GOPHERRC. To add a bookmark, simply type "a" when a menu item you want to recall is selected. That's a lower case "a" - type an upper case "A" to add the directory rather than the individual item as the bookmark. To use your bookmark file, just start Gopher with the command Gopher -b The initial screen will be a list of your recorded bookmarks. Naturally this will vary in a graphical interface.

Gopher Future

Many people expect Gopher to be replaced by various other tools based on a graphical interface and hypertext-like links. These have already been produced for several types of computer and can be much easier to use. Gopher will likely remain around for character-based terminals anyway.

WAIS

WAIS stands for "Wide Area Information System". It is a very similar system to Gopher, but with the distinction that full text searches are catered for more readily. In the same way as Gopher, WAIS consists of two parts. You run a client, which accesses servers that make information available. These are normally free although there are some fee-for-service servers. Client applications request documents using keywords. Servers search a full text index for appropriate documents and return a list of the documents containing the keyword. The client software may then request the server to send a copy of any of the documents found. Unfortunately few servers at present allow for compound (Boolean) searches, so there is a tendency to get either too few or too many hits. As well as providing a global service, WAIS can be used between a client and server on the same machine or a client and server on the same LAN, to provide a local information service for a single institution or company. Since the development of WWW, this has become the dominant use of WAIS.

Currently there are a large number of servers running with over 400 databases, ranging from recipes and movies to bibliographies, technical documents, and newsgroup archives. There seems to be more social science material accessible through WAIS than by any other means. Clients for WAIS are at an early stage in their development. Most often WAIS servers are accessed through Gopher menus, or from a Web browser. As an example, many bio-gophers have as a menu item a search of the WAIS archives for the BIONET/BIOSCI newsgroups/mailing lists.

World Wide Web

Like Gopher and WAIS, the World Wide Web, or just plain "web" or WWW, is a wide area information system. Unlike Gopher and WAIS, WWW is hypermedia based. So ubiquitous has WWW become (in a few months!) that it has almost become a synonym for Internet in some circles. To access the web, you run a local browser program. Like Gopher, the browser is a client, which accesses servers to fetch "documents". These may be standard text, graphical images, sounds, video, forms to fill in or one of a multitude of other document types. Information providers set up and publicise hypermedia servers from which browsers can get documents.

The browsers can, in addition, access files by FTP, NNTP (Network News Transport Protocol, the Usenet news protocol), Gopher and an ever-increasing range of other methods. On top of these, if the server has search capabilities, the browsers will permit searches of documents and databases. The documents that the browsers display are hypertext documents. Hypertext is text with embedded links to other text.

The browsers let you deal with the links transparently. Select the link and you are presented with the linked text regardless of where that text is located, whether on the same server or another.

In a production system, the web presents the user with a hypertext screen in which some phrases or icons are highlighted. Depending on the capabilities of the user's software (the "browser") the text may be highlighted either by being boxed, or coloured differently from the rest of the text. These highlighted regions are links, which are selected by clicking with a mouse, or by using arrow keys or however the browser interacts with the user. Documents stored on the server are written in a standardized markup language HTML (Hypertext Markup Language). Rather than storing emphasised text in, say, 12 point italic, it is written as plain text along with a label meaning "use whatever method to indicate emphasis is available". Your local computer can then format the text for your window size, screen abilities, even your printer. Converters can exchange between HTML and other document formats such as those used by popular wordprocessors.

If you lack a Web browser, you can get a taste using Telnet to connect to info.cern.ch (which is the European High-energy Physics Research Centre). This will automatically log you in and let you simulate hypertext links in text mode. (This site is a good source of pointers to subject information by the way). Many WWW clients are available. For graphical interfaces the most famous is Mosaic, or its later incarnation Netscape (which, in the author's opinion, is much preferable to Mosaic, even if not free to everyone). For text-only interfaces, the most commonly available client is Lynx, which is available for many flavours of Unix, for OpenVMS and probably other platforms.

WWW browsers integrate so many of the Internet services that they have become the first choice for many users, especially those using a PC or Macintosh. There are two important cautions. Firstly, downloading graphical images, a strength of WWW, will result in large bills for those who pay volume charges for Internet access, and may take a long time over slow links. Secondly, WWW pages are so easy to set up that there is a plethora of "Junk WWW" out there. Finding relevant and useful sites amongst them is not trivial.

Further Reading

Printed

There are many general books on Internet topics. Many become out of date very quickly, and few devote much space to topics of scientific interest. The following may be useful:

Browne, S. 1994. The Internet via Mosaic and World Wide Web. Ziff-Davis Press, Emeryville, CA.
Engst, A. 1994. Internet Starter Kit for Macintosh. Hayden Books, Indianapolis, IN,
Gaffin, A. 1994. Everybodys Guide to the Internet. The MIT Press, Cambridge, MA.
Krol, E. 1994. The Whole Internet User's Guide & Catalog 2nd Ed. M. Loukides (Ed.). O'Reilly & Associates, Sebastopol, CA.
LaQuey, T. and Ryder, J.C. 1993. The Internet Companion: A Beginner's Guide to Global Networking. Addison-Wesley, Reading, MA.

Raymond, E.S. 1991. The New Hacker's Dictionary. MIT Press, Cambridge, Mass.
Smith, R. 1994. Navigating the Internet. Sams Pub.

On-Line Documents

Documents available on-line may be more up to date than printed material. The locations given are not the only places these files may be found. Where possible, the home location is given.

de Presno, O. The Online World (A shareware book)
```
ftp://ftp.eunet.no/pub/text/online.txt
```

```
Hardy, H.E. 1993. The History of the Net.
umcc.umich.edu/pub/seraphim/doc/nethis8.txt
```

NCSA Education Group, The Incomplete Guide to the Internet.
available as a Microsoft Word document or as a Postscript file.
```
ftp://ftp.ncsa.uiuc.edu/Education/Education_Resources/
Incomplete_Guide/Dec_1993_Edition/MS_Word/*.hqx
```
(the document is in several files, for Macintosh)

```
ftp://ftp.ncsa.uiuc.edu/Education/Education_Resources/
Incomplete_Guide/Dec_1993_Edition/postscript/*.ps.Z
```
(several compressed Postscript files)

Yanoff, S. Speial Internet Connections
```
ftp://ftp.csd.uwm.edu/pub/inet.services.txt
```

Yanoff, S. Inter-Network Mail Guide. (How to mail that odd address from your odd address)
```
http://alpha.acast.nova.edu/cgi-bin/inmgq.pl
ftp://ftp.csd.uwm.edu/pub/internetwork-mail-guide
```

There are literally thousands of other sources; each of these lists many others.

From: *Internet for the Molecular Biologist.*
ISBN 1-898486-02-6 ©1996 Horizon Scientific Press, Wymondham, U.K.

4

CONNECTING

Gordon Findlay

Introduction

There are many ways of bridging the gap between your chair and the Internet. Getting access is generally not a problem. However, given the wide variety of ways in which it is possible to connect to the Internet, it pays to have some understanding of the options, and the strengths and weaknesses of each. There are two ingredients in the connectivity recipe: the "service provider" and the channel used to connect with the provider. While both are important; it is the combination of these which determines your capabilities, and the range of services available to you.

The Internet Provider

The means by which you will connect to the Internet, and the facilities available to you, will depend on the Internet Service Provider (ISP) you are using. Three somewhat overlapping types of enterprise will be able to offer you Internet access:

Academic or Scientific Institutions

These are the traditional suppliers. Obviously if you work in such an institution this is likely to be the most readily available, and probably the cheapest source. Most academic, scientific and research insitutions have some connection to the Internet. Not all offer the full range of the services mentioned in Chapter Three - for example, smaller institutions may lack the computing or communications power to take a full news feed. It may be possible to get a user account with a local University even if you are not a member of staff or a student.

Computer Hobbyist "Bulletin Boards"

Many of these bulletin boards have a form of connection with the Internet, often being only the ability to exchange Email with full Internet hosts, or read some part of Usenet News. These boards are often cheap or even free; but may have limited resources and only a small number of telephone lines.

Commercial "Internet Providers"

These are commercial organisations set up to make a profit by selling access to the Internet. The largest, such as America OnLine (AOL) and CompuServe, to name only two, have millions of subscribers. Typically these commercial providers offer a full range of services, which are charged for by a combination of the time spent connected and the volume of data moved. Many add value to their services by providing better indexing and organisation of Internet resources, although this indexing is more likely to extend to the home shopping market than molecular biology. Choosing a commercial Internet provider can become a research project in itself. There are several variables: the type of account offered; the costs involved in setting up and operating an account; the range of services available; the quality of the software offered and the quality of the service. Look carefully at how easy it is to actually connect. Many services find themselves short of telephone lines, especially during the early evening hours and in the weekends and school holidays. Ask colleagues and acquaintances for their experiences, and look out for "TUP" (Typical User Price) guides, which are sometimes distributed via news groups.

The Communications Channel

Just as there are three main groups of service providers, there are three principal ways of connecting a machine in your office (or den) to your provider. The traditional model is of a "dumb" terminal, connected by local cabling systems to a mainframe or minicomputer with Internet access. Very often these terminals are restricted to text-only, although X-windows terminals are not uncommon.

Most medium to large institutions have their own internal computer network, with one connection point to the Internet. The computer on your desk, whether it be UNIX, MSDOS, Windows or Macintosh, once connected to the local network can gain access to the Internet by running appropriate software. Connection to the local network requires an interface card in the computer, cabling to the network, and networking software (such as Novell, Windows for Workgroups, or Appletalk). This method of communication offers the widest range of software options: the desktop computer can emulate a text-based terminal, use a graphical interface, or run specially designed net access software.

Connections from home, or from institutions without a local network, are a little harder. The most economical approach is to connect the PC to a central site via the telephone lines and a modem. In some cases an ISDN connection may be an economical option. A modem is an electronic device which acts as an interface between the digital signals of the computer, and the analogue signals required by the telephone system. Modems are now quite inexpensive, especially the internal models, which fit in slots inside the computer box. Modems are capable of operating at different speeds. As usual, the faster the better, and 28,800 bps (bits per second) is very rapidly becoming the accepted standard. Until a few months ago 14,400 bps was the commonly accepted definition of a "fast" modem. If necessary, modems can operate at slower than their maximum speed, so if you are thinking of buying a modem, buy the fastest you can, even if the provider you are connecting to does not yet support the fastest speeds. Certainly graphical programs, such as web browsers, are less useful at slow speeds.

Of course an account on either an institution's or commercial provider's system is required together with the appropriate software. Dial up accounts may be "shell" accounts that provide an environment similar to a dumb terminal. More complex connections may be made using Point to Point Protocol (PPP) or Serial Line Internet Protocol (sl/IP) software. Obviously the host machine must support these protocols but where they are available the connection will allow the user access to a majority of the facilities provided by direct connections. The effectiveness of using PPP or sl/IP to connect to an ISP will depend on the speed of the link you are using. They will be best used with a high speed modem or an ISDN connection. PPP and sl/IP are functionally similar, the latter is a slightly older implementation and the two are not necessarily interchangeable. Some service providers may support both protocols but, as with all services, it is essential to find out what facilities any particular service provider offers.

Further Reading

Printed

There are many general books on Internet topics. Many become out of date very quickly, and few devote much space to topics of scientific interest. The following may be useful:

Browne, S. 1994. The Internet via Mosaic and World Wide Web. Ziff-Davis Press, Emeryville, CA.

Engst, A. 1994. Internet Starter Kit for Macintosh. Hayden Books, Indianapolis, IN.

Gaffin, A. 1994. Everybodys Guide to the Internet. MIT Press, Cambridge, MA.

Krol, E. 1994. The Whole Internet User's Guide & Catalog 2nd Ed. M. Loukides (Ed.). O'Reilly & Associates, Sebastopol, CA.

LaQuey, T. and Ryder, J.C. 1993. The Internet Companion: A Beginner's Guide to Global Networking. Addison-Wesley, Reading, MA.

Raymond, E.S. 1991. The New Hacker's Dictionary. MIT Press, Cambridge, Mass.

Smith, R. 1994. Navigating the Internet. Sams Pub.

On-Line Documents

Documents available on-line may be more up to date than printed material. The locations given are not the only places these files may be found. Where possible, the home location is given.

de Presno, O. The Online World (A shareware book)
`ftp://ftp.eunet.no/pub/text/online.txt`

Hardy, H.E. 1993. The History of the Net.
`umcc.umich.edu/pub/seraphim/doc/nethis8.txt`

NCSA Education Group, The Incomplete Guide to the Internet.
available as a Microsoft Word document or as a Postscript file.
`ftp://ftp.ncsa.uiuc.edu/Education/Education_Resources/`
`Incomplete_Guide/Dec_1993_Edition/MS_Word/*.hqx`
(the document is in several files, for Macintosh)

`ftp://ftp.ncsa.uiuc.edu/Education/Education_Resources/`
`Incomplete_Guide/Dec_1993_Edition/postscript/*.ps.Z`
(several compressed Postscript files)

Yanoff, S. Speial Internet Connections
`ftp://ftp.csd.uwm.edu/pub/inet.services.txt`

Yanoff, S. Inter-Network Mail Guide. (How to mail that odd address from your odd
address)
`http://alpha.acast.nova.edu/cgi-bin/inmgq.pl`
`ftp://ftp.csd.uwm.edu/pub/internetwork-mail-guide`

There are literally thousands of other sources; each of these lists many others.

From: *Internet for the Molecular Biologist.*
ISBN 1-898486-02-6 ©1996 Horizon Scientific Press, Wymondham, U.K.

5

SOFTWARE POINTERS

Gordon Findlay

Introduction

It is not possible to give a run-down here of even a small proportion of the software packages used by molecular biologists on the Internet. Rather, we point out some of the things which you should expect packages to do, include some advice about choosing between competing products and add a collection of folkloric observations.

Mailers

There are very, very many pieces of software around to use for sending mail. These mailers differ greatly. Some provide forms on screen for you to type in the name of the sender and recipient, with sophisticated editors to compose the message. Others adopt a more minimalist approach. Some know about the full range of Internet addresses; others need a little help.

Things to Look for

These are typical tasks which you should master:

* Sending a message to a local recipient (i.e. another local user).
* Sending a message to an Internet address.
* Replying to a message sent to you, incorporating the text and your comments, without retyping the whole thing.
* Adding, automatically, your signature block (".sig").
* Checking for new mail, deleting old messages no longer wanted, and filing important messages by topic (e.g. in "folders").
* Printing a mail message.
* Forwarding a copy of a message to someone else.

Many packages have a lot of options which allow you to customise the package to your use. At least be aware of the range of options available so that the package works for you, rather than vice versa.

Getting the Most out of Email

Like all tools, Email can be made more efficient by using it the right way:

- Use an editor. Always use an editor to compose messages. If your mailer doesn't have one built in, edit a file and send that.
- Keep copies. If a mail message is important, or you need to remember details, send a copy to yourself. However, don't do that for every trivial message; that soon becomes unwieldy.
- Holiday times. Vacation programs can be used to automatically send a reply saying something like "I'm not here but I'll be back... Talk to ... instead" so your correspondents don't wonder why you haven't replied (be careful with mailing lists though).
- Get mail in only one place. If you have accounts on multiple machines, or several accounts on a machine, get mail automatically forwarded from all your accounts to the account which you use most regularly and then you'll only have one place to check for mail.
- Be careful with enclosures. Some mailers allow you to send any files as attachments or as enclosures to messages. But Email is designed to handle simple text files only, not word processor documents, graphics files or whatever. Some mailers take special action to mail these sorts of files, but not all do. Don't assume that the person who receives your message will have a mailer which handles attachments in the same way.
- Don't forget to delete. Many users are reluctant to, or don't ever learn how to, delete old messages. None of the mailers are very good at indexing messages, and ploughing through lots of messages looking for the one you need is quite cumbersome. Of course, printing everything before deleting is another waste of resources.

News Readers

Obviously you need software to read News. Starting the software is as easy as typing NEWS, rn, inn or whatever it is called, or clicking the appropriate icon in a graphical environment. Generally you will select an initial set of groups which you wish to read. Thereafter these are the only groups which will be displayed to you unless you explicitly request others. There are two general ways to read news - read the new items, or read related items. A sequence of related items is called a "thread", and the better newsreaders allow you to read these related articles in order. News readers of course include the facilities to post new items, or comment on ("follow up") an item, and to respond to the poster of an item directly via Email. Many of the desirable features of mailers will be reflected here, and of course news readers and mailers can come in an integrated package.

Getting the Most out of News

Using News productively requires a little thought:

*   Read the replies, not the questions. Usually any reply to a question will quote the article posing the question, so you might save time by reading just the reply.
*   Your kill file is your friend. Most news reading software allows you to keep track of subjects or senders you don't want to bother with. Use this kill file freely.
*   Read the FAQ list first. Almost all groups have a number of questions which are asked time and again. These questions, and their answers are collected in the "FAQ" (Frequently Asked Questions) file, and it is advisable to read this file before posting a question. As well as being posted in their own group, the FAQs are posted in the group news.answers, and are available for anonymous FTP from the site rtfm.mit.edu, in the directories under /pub/usenet

FTP

Like most of the Internet protocols, FTP has been implemented in slightly different ways on different machines; the various sets of commands differ in detail. The command set is taken from the Unix operating system; even non-Unix implementations show this legacy. Graphical implementations are far easier to use than command-line version. The Unix origins of FTP are reflected in the case sensitivity of the commands and file names. The files foo.z and foo.Z are not the same.

Navigating

It can be very difficult to find a file you want in a large archive. If you are lucky you will know exactly where the file is (both site and directory). The name and location of a file are often given in this form:

```
garbo.uwasa.fi:/pc/sound/sound13.zip
```

which means that the file name is sound13.zip, in the directory /pc/sound on the site garbo.uwasa.fi. This is the traditional notation, although more references are now given in URL format. Using the URL format has become commonplace with the adoption of the WWW and is described later. If your FTP client is graphical, it will probably present a tree arrangement of directories and files. If you just get a textual list it can be difficult to tell which names on a list are files, and which are directories. In this case, using the "ls" command doesn't give you enough information, so try dir or "ls -l". One of them will probably give you a more informative display, indicating directories with "d" in the first column.

File Formats

File formats differ between machines in all sorts of ways. Even the form of a file name may differ: a.b.c is a legal file name in Unix, but not MSDOS. Even the very simplest files - straightforward lines of text - differ between operating systems, specifically in the way the end of a line is represented in the file. Therefore, transfering files between computers is fraught with potential problems. There are two modes, text mode and binary mode, in which transfers can be made. Text mode is used for "pure" text files only: files which are just lines of text. Binary mode is used for every other type of file. It is not always easy to anticipate which mode is appropriate, but word processor and spreadsheet files, executable programs, images and compressed archives (e.g. .ZIP files) must be transferred in binary mode. Postscript and ASCII files should be transferred in text mode, to avoid end of line problems.

Using Gopher

When using Gopher, you browse through the hierarchy of menus, or execute a search, to find the information you require. After viewing it (with whatever viewer is built in: they vary from machine to machine) you can save it in a (local) file, print it or perhaps Email it to yourself or someone else. Menu selections are made by either typing the number of the line you want, mouse clicking on the appropriate choice, or using the arrow keys to point at the line. Once the right line is highlighted, strike Enter, or double-click, to select the item. This item may be another menu, or a file, or an opportunity to enter a search statement. If it is a file, it will be presented on screen for you to view.

Most Gopher menus include an indication of how large a file is in kilobytes. Gopher downloads the whole file even if you only look at the first screen and discover the content is irrelevant. Therefore, avoid fetching a file which is very large, and irrelevant. This can cost time and money. From time to time you will temporarily leave Gopher, and enter a Telnet session, for example to search a library catalog. When you finish the session, you'll be returned to the point you were at in the Gopher menus. It is important to remember the instructions that Gopher gives you at the start of a Telnet session, such as the name to use to log in. These instructions usually disappear from the screen just as you are about to use them! Like the other tools, Gophers vary between systems: a Macintosh Gopher client is designed to use the standard Macintosh interface for example. For this reason we cannot give explicit instructions here. However, all Gophers have built in help.

WEB Browsers

As noted in Chapter 3, using a Web browser is a good way of getting access to a large number of the available Internet services all in one convenient package. Although there are a fair number of browsers, the most common by far is the graphical browser Netscape (Figure 1). Netscape was developed as an alternative to an earlier graphical browser named Mosaic. If you still have Mosaic you should see if it is possible to upgrade to Netscape as it has many improvements over Mosaic. A most important status line appears at the foot of the screen. This indicates what is happening, such as a site being contacted, or a document being retrieved. Note that the size of a download might not be obvious before it starts; a progress indicator appears at the bottom right. The "Stop" icon may be used to interrupt a transfer which is either unexpectedly long, or making unacceptably slow progress.

Figure 1. An example of Netscape's appearance. The figure shows the Macintosh version of the program configured to display all possible tool bars, icons and status bar indicators. Most of these may be turned off in the program preferences. The figure has been constructed to show all the tools as they appear when active, they are often not all active at the same time. Inactive tools are usually greyed. Versions for different computer platforms are very similar with minor variations depending on the operating system.

Installation

Whatever browser you use, expect to spend some time installing and configuring the software. Typically you need not only the browser software itself, but

- A TCP/IP stack. A set of programs that allow your computer to use TCP/IP to communicate down whatever channel you are using.
- Client programs which the browser uses to access particular services, such as a News reader or a Telnet client. These are used for the services which are not built-in to the browser software itself.

Try to get a complete set of utilities from one place: either a shareware collection or commercial product. If you are using a dial-up connection (called sl/IP or PPP) you will be well served by sticking to a set provided, or at least recommended, by your ISP.

Configuration Issues

Any tool as flexible as a graphical WWW browser has a lot of configuration options. You'll need to configure (or have configured for you) such items as:

- Your Email address, personal name, and signature file.
- The system which your computer uses as a "smart mailer". (A "smart mailer" is a

system that your computer passes mail to, and which is assumed to be a bigger player in the Email world, handling issues such as mail routing for you. Typically this will be a Unix or VMS system operated by your Internet Service provider.)
• The system to obtain news feed from.

Individual preferences are catered for, you may select:

• The colours and sizes of screen windows.
• The way in which links are indicated on screen.
• Whether to indicate in a different way the links that you have visited.

Most of these issues are dealt with on a "Preferences" menu, although in Mosaic it may be necessary to edit the file MOSIAC.INI manually.

Home Page and Graphics loading

Two configuration options are especially important. The "home page" is the page to which the browser connects when started. For example, the default home page for Netscape is found at the site

WWW.NETSCAPE.COM.

Set your home page either to a convenient site which you refer to often and has lots of useful links, or to a blank page, so that nothing is loaded until you open a location by name. This can save a good bit of time, and communication bandwidth. Once you become familiar with the software, you can build your own home page, with links to all the most important (to you) sites. There are many candidates in the later chapters of this book.
        The graphical browsers can, of course, download graphics, and most Web pages have graphic images embedded in them. However, downloading graphics takes much more time than downloading text, and very often the graphics turn out to be irrelevant. Most experienced users choose to configure the software so that graphics are not downloaded unless requested. When a graphic is present a "placeholder" appears; click on the placeholder to download the actual image. Again, time and communication costs are saved by not downloading unwanted pictures.

WWW Locations and the URL

Everything is known as a "location" on the Web, whether it be a hypertext page, news group or interactive login (telnet session). Locations are addressed by using their Universal Resource Locator, or URL. While Web browsing or "surfing" is most enjoyable, more work gets done by opening a location by typing its URL directly. "Open a location" is usually on the file menu of Web browsers. URLs include the machine name, type of access and sometimes a file name. The access type, whether it be hypertext, gopher or other is followed by a colon, and two slashes to separate it from the machine name. The machine name may be followed by a file and, occasionally, additional information. Here are some real examples, and commentary:

`http://anaeswww.chmeds.ac.nz/chch.html`
- a hypertext source; the machine name is `anaeswww.chmeds.ac.nz`, and the file name of the first page to load is `chch.html`.

`gopher://bdt.ftpt.br`
- a gopher server in Brazil

`telnet://jhuvm.hcf.jhu.edu`
- a telnet location. What is not given in the URL is the username and password required to log in. These are usually included in the commentary surrounding a link.

`ftp://ftp.lib.muohio.edu/pub/net.workshop`
- an FTP site, and a file name.

Unfortunately URLs become quite long, and are case-sensitive, presenting a challenge to two-finger typists.

Tracking the Good Stuff

Apart from customising your own home page, which requires some understanding of the Hypertext Mark-up Language (HTML), there are two ways of keeping track of the sites and resources you have visited. Browsers generally keep a list of the most recent links used: the history list. To revisit, "open" the history list and click (or double-click) on the item. History lists are limited in size, and once full, each addition bumps the oldest current entry off. Bookmarks are more versatile. You may add as many as you wish to. These are saved in an external file. Add bookmarks as you find important sites. A good bookmark list can be the heart of a customised home page.

Saving Pages and Data

Often you will want to save information you have downloaded. Files may be saved either in straight text ("ASCII") format, or in source format: the actual HTML describing the page. Saving interesting or appealing pages in HTML will help if you ever wish to learn to set up your own pages. When downloading a file of data (by some form of ftp, usually built in) you may be asked for the name of a "viewer". This means that the browser software doesn't recognise the file type. Usually just choose to treat the file as a "byte stream on disk" is all that is required.

Acknowledgement

Parts of Chapters 1 - 5 are based (with permission) on material from:

Findlay, G. and Dawson, L. 1994. The Well Connected Library. Wellington, NZ. New Zealand Library and Information Association.

Further Reading

Printed

There are many general books on Internet topics. Many become out of date very quickly, and few devote much space to topics of scientific interest. The following may be useful:

Browne, S. 1994. The Internet via Mosaic and World Wide Web. Ziff-Davis Press, Emeryville, CA.

Engst, A. 1994. Internet Starter Kit for Macintosh. Hayden Books, Indianapolis, IN.

Gaffin, A. 1994. Everybodys Guide to the Internet. MIT Press, Cambridge, MA.

Krol, E. 1994. The Whole Internet User's Guide & Catalog 2nd Ed. M. Loukides (Ed.). O'Reilly & Associates, Sebastopol, CA.

LaQuey, T. and Ryder, J.C. 1993. The Internet Companion: A Beginner's Guide to Global Networking. Addison-Wesley, Reading, MA.

Raymond, E.S. 1991. The New Hacker's Dictionary. MIT Press, Cambridge, Mass.

Smith, R. 1994. Navigating the Internet. Sams Pub.

On-Line Documents

Documents available on-line may be more up to date than printed material. The locations given are not the only places these files may be found. Where possible, the home location is given.

de Presno, O. The Online World (A shareware book)
`ftp://ftp.eunet.no/pub/text/online.txt`

Hardy, H.E. 1993. The History of the Net.
`umcc.umich.edu/pub/seraphim/doc/nethis8.txt`

NCSA Education Group, The Incomplete Guide to the Internet.
available as a Microsoft Word document or as a Postscript file.
`ftp://ftp.ncsa.uiuc.edu/Education/Education_Resources/`
`Incomplete_Guide/Dec_1993_Edition/MS_Word/*.hqx`
(the document is in several files, for Macintosh)

`ftp://ftp.ncsa.uiuc.edu/Education/Education_Resources/`
`Incomplete_Guide/Dec_1993_Edition/postscript/*.ps.Z`
(several compressed Postscript files)

Yanoff, S. Speial Internet Connections
`ftp://ftp.csd.uwm.edu/pub/inet.services.txt`

Yanoff, S. Inter-Network Mail Guide. (How to mail that odd address from your odd address)

`http://alpha.acast.nova.edu/cgi-bin/inmgq.pl`
`ftp://ftp.csd.uwm.edu/pub/internetwork-mail-guide`

From: *Internet for the Molecular Biologist.*
ISBN 1-898486-02-6 ©1996 Horizon Scientific Press, Wymondham, U.K.

6

CONTRIBUTING TO THE INTERNET

Gordon Findlay

Introduction

Publishing your own data, software and ideas using the Internet can be a very satisfying experience, or a very frustrating one. Satisfying if you succeed in distributing worthwhile material to people who can make use of it; frustrating if you spend a lot of your time fiddling with computer and communication problems. If you have resources you wish to share, there are a few questions which you must answer before you determine how to go about it:

- How shall the resources be distributed? By anonymous FTP? Via Gopher, or via Email to requesters? Using WWW pages?
- Where should you host the resources? Your own institutional system? Or would it be preferable to have your material distributed through one of the established biomedical Internet sites?
- How will the resources be made known?
- How often will the the material be updated? How can you track users of your resources, to advise them of updates?
- Where will you obtain the communication and computer expertise required?
- How much time have you to devote to maintaining these resources, answering questions about them and solving the problems which are sure to arise once other scientists start to use them?
- What will happen if (when?) you get that fabulous new job in another place, or change your research interests?

How to Publish on the Internet?

The way in which material is distributed depends on both the type of material, and its volume. Later sections of the book will list many sites, and means to access them. Look at a variety of them, and consider the approach that they use. How easy is it to use the various forms of access provided at each site? Would the approach used by this site suit your data?

Undoubtedly the easiest way to distribute files, whether data or software, is by anonymous FTP. Setting up an anonymous FTP site just means finding some disk space, creating a directory and user name, and moving your files into the right place.

There are some security issues, which are resolved by setting file and directory rights correctly. Be aware that some WWW browsers, notably many versions of Mosaic, cannot retrieve FTP files from non-Unix sites. Hopefully, this is a short-lived problem. If many files are to be made available consider using Gopher as the medium, with perhaps some simple searches to aid file selection. Setting up a Gopher server is not a straight-forward task; the software can be difficult to configure. Writing the link files is not transparent, and requires detailed planning. On the other hand, Gopher is an easy way to access the data. This illustrates the trade-off; the easier it is for your users to find and retrieve data, the more work is required to configure the server.

Of course the most up-market approach is to use WWW. Writing WWW pages is not a straight-forward exercise; some mastery of HTML is required. Be aware that there are different versions of HTML; some browsers and some servers are more restrictive than others in what they can handle. At the time of writing, a lot of work is being done on some extensions to HTML, particularly in relation to printing. Look at the source HTML of pages which show the sort of approach you wish to use, and study the way that the various HTML constructs are reflected in what you see in your browser. WWW pages are a great way to distribute graphics; and to provide a semi-interactive environment, using feedback and forms as requesters. Be sure to look at the pages in more than one browser, and for completeness, see what they look like in a text-only browser, such as Lynx. Some people, for financial and communication reasons, will be restricted to character- cell terminals for some time yet. Many commercial internet service providers will, for a fee, write WWW pages for you. Of course, WWW pages may need to be changed frequently, making this a poor option for many.

Where to Publish?

You probably want your data to be available 24 hours a day. This means that the system you are distributing through must be available, and connected to the Internet, most of the time. This does not mean that a large system is required: it is possible to operate a WWW server on a low-powered Macintosh for example. However, if your institution is already running a server of some sort you should try to make use of it. Consider if logging of users is required. If so, decide what level of logging is needed and choose a hardware platform that can provide these requirements. For example, Unix-based WWW servers offer far more transaction logging and security features than do present MSDOS or Macintosh servers.

Publicising your Work

Announcements of the availability of new and updated resources frequently appear in news groups and mailing lists. If the data you wish to make available is of widespread interest, arrange for links to it to be placed on major Gopher and WWW servers. If others are also distributing copies of your material, be careful that any pointers to these additional copies reference the prime site. Word of mouth is a prime means for spreading information on the Internet. It is important that users who find a copy of your material are able to check that it is indeed current.

Administering Mailing Lists

There is a real temptation to set up mailing lists at the drop of a hat! They can be a useful tool for keeping a group of workers in some field in touch with each other. Mailing lists would be easy to run if mailing addresses, and users' mailer software, were even slightly more standardised. As it is, there are so many variations that automatic list software will generate plenty of anomalies and errors for manual handling. If you do wish to set up a mailing lists, contact your system manager without delay. Unless he can give you access to properly installed and configured software, such as Majordomo, Listserv or Maiser, running a mailing list will require a lot of tedious and often difficult manual work. The Biosci/Bionet administrators may be prepared to trial a mailing list on your behalf.

Distributing Software

Distributing software is fraught with problems. Most system managers will want source code, not just compiled executable programs, for several reasons, including the avoidance of viruses and worms, and the need to modify the code for local conditions. Operating system upgrades frequently mean that code must be recompiled as well. There are two approaches to developing software; either make it fancy, or make it portable. Given the very wide range of systems presently in use in molecular biology, it is impossible to provide versions (ports) for everyone. Encourage people who do convert your code to their systems to send their ports back to you. There are several well known repositories for public-domain and shareware software. The European Molecular Biology Laboratories (EMBL) and Indiana University are two of the most important.

Further Reading

Printed

There are many general books on Internet topics. Many become out of date very quickly, and few devote much space to topics of scientific interest. The following may be useful:

Browne, S. 1994. The Internet via Mosaic and World Wide Web. Ziff-Davis Press, Emeryville, CA.

Engst, A. 1994. Internet Starter Kit for Macintosh. Hayden Books, Indianapolis, IN.

Gaffin, A. 1994. Everybodys Guide to the Internet. MIT Press, Cambridge, MA.

Krol, E. 1994. The Whole Internet User's Guide & Catalog 2nd Ed. M. Loukides (Ed.). O'Reilly & Associates, Sebastopol, CA.

LaQuey, T. and Ryder, J.C. 1993. The Internet Companion: A Beginner's Guide to Global Networking. Addison-Wesley, Reading, MA.

Raymond, E.S. 1991. The New Hacker's Dictionary. MIT Press, Cambridge, Mass.

Smith, R. 1994. Navigating the Internet. Sams Pub.

On-Line Documents

Documents available on-line may be more up to date than printed material. The locations given are not the only places these files may be found. Where possible, the home location is given.

de Presno, O. The Online World (A shareware book)
`ftp://ftp.eunet.no/pub/text/online.txt`

Hardy, H.E. 1993. The History of the Net.
`umcc.umich.edu/pub/seraphim/doc/nethis8.txt`

NCSA Education Group, The Incomplete Guide to the Internet.
available as a Microsoft Word document or as a Postscript file.
`ftp://ftp.ncsa.uiuc.edu/Education/Education_Resources/`
`Incomplete_Guide/Dec_1993_Edition/MS_Word/*.hqx`
(the document is in several files, for Macintosh)

`ftp://ftp.ncsa.uiuc.edu/Education/Education_Resources/`
`Incomplete_Guide/Dec_1993_Edition/postscript/*.ps.Z`
(several compressed Postscript files)

Yanoff, S. Speial Internet Connections
`ftp://ftp.csd.uwm.edu/pub/inet.services.txt`

Yanoff, S. Inter-Network Mail Guide. (How to mail that odd address from your odd address)
`http://alpha.acast.nova.edu/cgi-bin/inmgq.pl`
`ftp://ftp.csd.uwm.edu/pub/internetwork-mail-guide`

There are literally thousands of other sources; each of these lists many others.

From: *Internet for the Molecular Biologist.*
ISBN 1-898486-02-6 ©1996 Horizon Scientific Press, Wymondham, U.K.

7

SEQUENCE RETRIEVAL AND ANALYSIS USING ELECTRONIC MAIL SERVERS

T. S. Pillay

Introduction

This chapter provides an introduction to the various Email servers available to the molecular biologist in as far as they pertain to sequence analysis. The reader is particularly referred to the summary table of Email servers which may be copied and pinned up next to the computer terminal (Table 1). From this readers can select the server most suited to their needs. Although the most popular servers are discussed in some length (e.g.. Retrieve and Blast), a discussion of the services provided by every server is beyond the scope of this chapter and detailed documentation is readily and rapidly available by sending a "help" message to the electronic mail address listed in the summary table. If, for any reason, communication with the server is problematic then an address to report problems is also provided in Table 1. The more advanced reader is referred to the list of references and further reading provided at the end of the chapter.

Sequence analysis by Email server is useful for scientists who do not have access to onsite mainframe computers or expensive software for doing sequence analysis. These involve obtaining an account with the mainframe and the user is charged a fee according to the amount of CPU time used. Most individuals at universities have free access to electronic mail and sequence analysis by electronic mail removes this impediment. Furthermore, the user has access to a number of servers and databases which may offer slightly different services. Email servers also use more powerful hardware and consequently more sophisticated sequence analysis is possible than that available on a personal computer. The only disadvantage and a minor one, is that the request gets queued and at busy times the user may have to wait several hours for the results of an Email search or Email analysis.

There are probably two main tasks which users may wish to perform:

a)   Obtaining a sequence
b)   Analysing it in detail to search for motifs and similarities with other proteins.

The sequence may initially be in the users possession as a result of a cloning project. In this instance, the first task will be to determine if the sequence exists in the various database with a view to answering the question of whether the gene for this protein has

Table 1. Email Servers at a Glance

Server name	E-mail address	Retrieval	Similarity searching	Gene prediction	Protein analysis	Address to report problems
BIOSERVE	bioserve@temin.lanl.gov	+				michael@genome.lanl.gov
EBI NETWORK FILE SERVER	netserv@ebi.ac.uk	+				nethelp@ebi.ac.uk
AFLAT DB NETWORK SERVER	flat-netserv@smlab.eg.gunma-u.ac.jp					smiyazaw@smlab.eg.gunma-u.ac.jp
GDB/OMIM MAIL SERVER	mailserv@gdb.org	+	+			help@gdb.org
GENIUSnet	netserv@genius.embnet.dkfz-heidelberg.de	+				dok419@genius.embnet.dkfz-heidelberg.de
PIR NETWORK						
REQUEST SERVICE	fileserv@nbrf.georgetown.edu	+	+			postmaster@nbrf.georgetown.edu
PDB Email file server	fileserv@pb1.pdb.bnl.gov	+				skora@bnl.gov
RETRIEVE	retrieve@ncbi.nlm.nih.gov	+				retrieve-help@ncbi.nlm.nih.gov
	retrieve@ebi.ac.uk					retrieve-help@ncbi.nlm.nih.gov
RETRIEVE dbEST Server	retrieve@ncbi.nlm.nih.gov	+				retrieve-help@ncbi.nlm.nih.gov
RETRIEVE dbSTS Server	retrieve@ncbi.nlm.nih.gov	+				vriend@embl-heidelberg.de
TM7						gene-server-management@evolution.bchs.uh.edu
University of Houston Gene Server	gene-server@bchs.uh.edu	+				lsestern@weizmann.weizmann.ac.il
BICSERV	bicserv@sgbcd.weizmann.ac.il		+			bioscan-info@cs.unc.edu
BIOSCAN	bioscan@cs.unc.edu		+			blast-help@ncbi.nlm.nih.gov
BLAST	blast@ncbi.nlm.nih.gov		+			nethelp@ebi.ac.uk
BLITZ	blitz@ebi.ac.uk		+			henikoff@howard.fhcrc.org
BLOCKS	blocks@howard.fhcrc.org		+			
DAPMAIL	dapmail@ed.ac.uk		+			knecht@inf.ethz.ch
DARWIN	cbrg@inf.ethz.ch		+			reczko@dkfz-heidelberg.de
DEF mailserver	def@mbp-sgi4.inet.dkfz-heidelberg.de				+	
DFLASH	dflash@watson.ibm.com		+			domain-comment@hubi.abc.hu
DOMAIN	domain@hubi.abc.hu		+			

Table 1. Email Servers at a Glance (continued)

Server name	E-mail address	Retrieval	Similarity searching	Gene prediction	Protein analysis	Address to report problems
EBI Mail-FASTA	fasta@ebi.ac.uk		+			nethelp@ebi.ac.uk
EBI Mail-PROSITE	prosite@ebi.ac.uk		+			nethelp@ebi.ac.uk
EBI Mail-QUICKSEARCH	quick@ebi.ac.uk		+			nethelp@ebi.ac.uk
EMBL Mail-FASTA	Fasta@ebi.ac.uk Fasta@EMBL-Heidelberg.de		+			NetHelp@ebi.ac.uk
GCRDb Email FASTA	gcrdb@receptor.mgh.harvard.edu		+			lfk@receptor.mgh.harvard.edu
GENIUSnet MAIL-FASTA	mfasta@genius.embnet.dkfz-heidelberg.de		+			dok419@genius.embnet.dkfz-heidelberg.de
GenQuest	q@orml.gov		+			grailmail@orml.gov
PYTHIA	pythia@anl.gov			+		pythia-admin@anl.gov
RDP mailserver	server@rdp.life.uiuc.edu	+				rdp@phylo.life.uiuc.edu
SBASE	sbase@icgeb.trieste.it		+			sbase-comment@icgeb.trieste.it
GENEFINDER	service@bchs.uh.edu			+		solovyev@cmb.bcm.tmc.edu
GENEID	geneid@bir.cedb.uwf.edu			+		
GRAIL	grail@orml.gov			+	+	grailmail@orml.gov
HUGEMAP	hugemap@genethon.fr			+		
NETGENE	netgene@virus.fki.dth.dk			+		engel@virus.fki.dth.dk
GENMARK	genmark@ford.gatech.edu			+	+	mb56@prism.gatech.edu
PREDICTPROTEIN	predictprotein@embl-heidelberg.de				+	predict-help@embl.heidelberg.de
Motif	motif@genome.ad.jp		+			motif-manager@genome.ad.jp
MOWSE	mowse@dl.ac.uk				+	mbdpn@s-crim1.dl.ac.uk
nnPREDICT	nnpredict@celeste.ucsf.edu				+	nn-predict-request@celeste.ucsf.edu
Prodom	prodom@toulouse.inra.fr		+		+	proquest@toulouse.inra.fr
PSORT	psort@nibb.ac.jp				+	nakai@nibb.ac.jp
SOPM	deleage@ibcp.fr				+	deleage@ibcp.fr
TMAP	tmap@embl-heidelberg.de				+	persson@embl-heidelberg.de

been cloned and sequenced before. This will be dealt with later in the chapter. There are more than 40 Email servers that are able to provide a wide variety of sequence analysis functions ranging from simple database search and retrieval to powerful sequence analysis and comparison. This essentially means a user sitting at home or in the lab equipped with a personal computer and a modem can have access to a worldwide assortment of extremely sophisticated sequence analysis functions controlled by simple Email text messages. The Electronic mail servers are computers programmed to respond in a specific and automated way to text messages received in a rigidly defined format. The message sent to the server is in essence a set of commands in a specific format defined by each server. Any deviation from this causes the server to generate a programmed response usually consisting of mailing the "help" file for that server.

The first step in sequence analysis usually involves actually extracting the sequence in a computer-readable format for a variety of purposes including restriction analysis, primer and oligonucleotide designs. Published sequences are usually submitted to a particular database and the sequence is given a unique identifier, the accession number, which makes it possible to extract the sequence by this identifier alone. Not having an accession number makes the task a little more difficult as one is then compelled to search by using a keyword, authors name, citation etc.

Obtaining an Amino Acid or Nucleotide Sequence by Email

Databases available for sequence retrieval and Email addresses:

1. Retrieve — retrieve@ncbi.nlm.nih.gov
2. Bioserve — bioserve@temin.lanl.gov
3. Network File Server — netserv@ebi.ac.uk
4. GENIUSnet sequence-server — netserv@genius.embnet.dkfz-heidelberg.de
5. Flat DB Email Network Server — flat-netserv@smlab.e.g. gunma-u.ac.jp
6. PIR Network Request Service — fileserv@nbrf.georgetown.edu

Note that detailed documentation for each of these servers can be obtained by sending the message "help" to these addresses.

The Retrieve Server

The "Retrieve Server" is maintained by the National Center for Biotechnology Information (NCBI), National Library of Medicine, NIH, in Bethesda, Maryland. The Internet address is retrieve@ncbi.nlm.nih.gov. One can search several sequence databases although each has to be searched with a separate message. Sequences are obtained from the Retrieve Server in the following way: Compose an Email message and send it to the Internet address:

retrieve@ncbi.nlm.nih.gov

The message should read as follows (if you want to search Genbank and using the SHC protein as an example):

```
Datalib Genbank
maxdocs   3
Begin
SHC
```

Explanation of Email Message

Datalib Genbank	Mandatory. The Datalib command asks the server to search a specified database, in this case Genbank (see Table 2). Only one database can be searched with each message.
Maxdocs 3	This specifies the number of matching documents the server should transmit back to the sender. Maxdocs is one of several commands that may used to modify the search output, details of which can be found in the help file from retrieve. The default is 20 and the limit is 2400. This and other commands allows the user to limit the amount of information received.
Begin	Mandatory. This marks the end of the commands. The next part of the message is to be used for the search.
SHC	This is the search "phrase". Boolean operators such as AND, NOT, OR can be used. This can be a keyword or an accession number.

Using Accession Numbers

```
Datalib Genbank
maxdocs   3
Begin
X68148
U15784
```

X68148	Accession number. More than one accession number can be included in a message

A number of databases can be searched using Retrieve (see Table 2 for abbreviations). There are other optional parameters which can be included in the message sent to the Retrieve server. DATALIB and BEGIN are the only mandatory commands.

MAXDOCS	This sets a limit to the maximum number of documents to transmit. Default is 20. The maximum is 2400.
MAXLINES	This sets a limit to the number of lines to transmit of each document. Default is 1000. The maximum is 50 000
TITLES	If the user only wants to retrieve titles of each document, this should be used.

Table 2. Database Abbreviations used by Different Servers

Server	Database	Abbreviation
Retrieve	Genbank	gb
	Genbank update	gbu
	Genbank full release	gbonly
	EMBL DNA	embl or emb
	EMBL update	emblu
	EMBL Full release	emblonly
	Swiss-Prot	sp
	Swiss-Prot updates	spu
	PIR	pir
	Vector subset of Genbank	vector
	Vecbase	vecbase
	GenPept	gp
	GenPept update	gpu
	Kabat's database	kabatnuc
	Kabat's database-protein	kabatpro
	Eukaryotic promoter database	epd
	Protein Data Bank	pdb
	Transcription Factors	tfd
Flat DB	Genbank	gb
	EMBL	embl
	Translated Genbank	gp
	Swiss-Prot	swiss
	PIR Protein database	pir
	Protein Research Foundation	prf
Blast	Genbank to last major release	gb
	Genbank update to last major release	gbupdate
	EMBL Data library to last update	emb
	3D Brookhaven Protein Databank	pdb
	Alu repeats from REPBASE	alu
	Vector subset of GenBank, NCBI	vector
	Kabat's database of sequences of immunological interest	kabat
	Expressed Sequence tags (ESTS)	dbest
	Sequence Tagged sites (STS)	dbsts
	Eukaryotic Promotor database	ep
	Swiss-Prot, PIR(R), GenPept and GenPept updates	nr
	Swiss-Prot	sp
	Last major release of NBRF PIR(r)	pir
	Swiss-Prot update	sp
	Translated Genbank and updates	gp
	Brookhaven PDB	pdb
	Kabat's database	kabat
	Transcription factors database	tfd
	Ancient Conserved region subset of Swiss-Prot	acr

STARTDOC Starting document number. This is used in successive mail messages when more data is to be received than previously allowed by the maximum values of MAXDOCS and MAXLINES.

PATH Return address if problems with delivery are anticipated or if the documents should be sent to another address.

Queries can also be modified by restricting the search to particular fields of records.

For example, using the search query above:

```
Datalib Genbank
maxdocs  3
Begin
SHC [key]
```

Adding the field abbreviation [key] asks the server to only retrieve records with the word SHC in the "keyword" field. A complete list of the field abbreviations is listed in the retrieve help file.

The EBI Server

The EBI sequence server is used by sending GET commands to the address netserv@genius.embnet.dkfz-heidelberg.de. It should be noted that although the servers EBI Network File Server (netserv@ebi.ac.uk) and GENIUSnet sequence-server (netserv@genius.embnet.dkfz-heidelberg.de) are listed as different servers, they are both part of the EMBL network. Multiple GET commands can be sent in the same message. To obtain the sequences SHC (from SwissProt) and the sequence with the accession number X68148 (EMBL) the message should read as follows:

```
GET PROT: SHC
GET NUC: X68148
```

PROT and NUC specify that the user wants a protein and nucleotide sequence respectively. The following databases are available, there is no need to specify the database in the Email message:

EmNew	latest EMBL entries
EMBL	nucleotide sequences
NBRF/Pir	protein sequences
SwissProt	protein sequences

The Flat DB Email Network Server

The Flat Server is located at Gunma University, Faculty of Technology and uses the SCANDB command in Email messages. The following databases are available (see Table 2 for abbreviations):

Genbank	All Genbank releases and new entries
EMBL	All EMBL releases and new entries
Genpept	Translated Genbank
Swiss	SwissProt protein database
PIR	PIR protein database
PRF	Peptide Research Foundation peptide database

Examples of the format of the message:

```
scandb gb 'HSSHC'
scandb gb -a 'X68148'
```

scandb gb 'HSSHC'	This asks the server to scan the Genbank database for records with this entry name.
scandb gb -a 'X68148'	This asks the server to scan the Genbank database for entries with this accession number.

Determining if Identical or Similar Sequences Exist in the Sequence Databases

Commonly, a molecular biologist obtains a potentially new sequence from a cloning project. The task now faced is how to determine if the sequence exists in the databases or if there are any sequence similarities with other sequences in the database. A number of Email servers are available to address this question. All are fully automated and use various algorithms to search either all the databases or the specified database based on the commands the user sends to the server.

Servers for sequence similarity database searching:

1.	BLAST	`blast@ncbi.nlm.nih.gov`
2.	BLITZ	`blitz@ebi.ac.uk`
3.	EBI Mail-FASTA	`fasta@ebi.ac.uk`
	GENIUSnet Mail-FASTA	`mfasta@genius.embnet.`
		`dkfz-heidelberg.de`
4.	QUICK	`quick@ebi.ac.uk`
5.	Flat DB Email Network Server	`flat-netserv@smlab.e.g..`
		`gunma-u.ac.jp`
6.	Bicserv	`bicserv@sgbcd.`
		`weizmann.ac.il`
7.	BioSCAN	`bioscan@cs.unc.edu`
8.	dFLASH	`dflash@watson.ibm.com`
9.	GenQuest	`q@ornl.gov`

A number of other servers are also listed in the summary along with those for searching for protein motifs and patterns. For the new user, the author would recommend starting with the EBI Mail QUICKSEARCH server (`quick@ebi.ac.uk`). This server is the easiest to use and interpret. The server uses an implementation of the QUICKSEARCH and QUICKSHOW programs (Devereux *et al.*, 1984). More detailed analysis can be performed using the EMBL FASTA, BLAST and BLITZ servers.

Using QUICKSEARCH (nucleotides only)

Send a message to `quick@ebi.ac.uk` or `quick@embl-heidleberg.de` in the following format:

```
Title   Clone1
Seq
agctgcatagatcgtcgatacgtacgtagtc......
End
```

Title   Subject line of the message.

Seq   is the only mandatory command. Everything following "Seq" is treated as part of the sequence until "end" is reached or the end of the mail message.

Note some Email programs insert the users signature automatically-hence adding "end" is a good idea. Each command should be on a separate line. The server will use default values for other commands unless the user overrides them by adding modified values. These commands are:

LIBRARY
    All; all EMBL and Genbank entries
    Genew; new EMBL and Genbank entries

WINDOW -STRINGENCY
    Window and Stirngency alter the sensitivity of the search. Default values are calculated from the length of the sequence. Decreasing Window and Stringency will increase the sensitivity of the search.

PERFECT
    This will ask the server to report only perfect matches.

MATCH
    Specifies the percentage overlap of identity required. Default is 90.

BEST
    Specifies the algorithm to use. Default is a Needleman-Wunsch alignment (Needleman and Wunsch, 1970). Adding Best asks the server to use the Smith-Waterman algorithm (Smith and Waterman, 1981).

ONE
    Adding this asks the server to search only the strand given and notthe complementary strand.

The EMBL FASTA Server

Address   Fasta@ebi.ac.uk
              Fasta@EMBL-Heidelberg.de

The EMBL FASTA Server uses Mail-FastA which is based on the FASTA program (Pearson and Lipman, 1988). This allows fast and sensitive comparisons of nucleic acid or protein sequences in a situation when the user wishes to determine if a related sequence can be found in the following databases:

Nucleotide: EMBL, Genbank and latest releases and new entries.
Protein: Swiss-Prot, PIR/NBRF and latest releases and new entries; Brookhaven (PDB) structure database.

While FASTA will detect related nucleotide or protein sequences, a faster search may be performed using QUICKSEARCH (Quick@ebi.ac.uk). QUICKSEARCH is less demanding of CPU time and will serve most purposes. QUICKSEARCH was developed for rapid database searching and is useful for determining if a DNA sequence or a closely related sequence exists in the database. More distantly related DNA sequences should be searched for using FASTA or BLAST. Furthermore QUICKSEARCH cannot be used for protein sequences. The format of the Email message sent to the FASTA server is very similar to that used for QUICKSEARCH with a few additional commands.

Example of a FASTA message:

```
Title test gene
List 15
Align 3
One
Seq
agctgcatagatcgtcgatacgtacgtagtc
End
```

Title test gene	Subject line of message.
List 15	Number of top scoring results to list.
Align 3	This asks the server to take the three best scoring sequences and align them.
One	This asks the server to only use the strand given in the message.
Seq	Mandatory, signifies that the following text is the sequence to use for the search.
agc...	This is the search sequence.
End	End of message.

The following are optional parameters:

LIB	Library name (default is EMALL - all EMBL entries and SWALL-all Swiss-Prot entries).
WORD n	n is between 1 and 6. 6 is the default. The smaller the value the higher the sensitivity of the search but small values increase the search time dramatically.
LIST n	This instructs the server to list only the "n" top scoring results. Default is 50. Maximum is 100.
ALIGN n	This asks the server to align the "n" best scoring sequences. Default is 10 and the maximum is 30.
PROT	For very small protein sequences use PROT to force FASTA to treat the sequence as a protein sequence.

The BLAST (Basic Local Alignment Search Tool) Email Server

BLAST is used to compare a nucleotide or amino acid query against the various databases. The programs use algorithms for finding ungapped, locally optimal sequence alignments. This is usually useful for the molecular biologist who has cloned and partially sequenced a particular cDNA and now wants to find out if it exists in the databases already. The user performs the similarity search by sending the server a message in the format detailed below and which contains the nucleotide or protein sequence. The server then compares the sequence against the database specified and returns the results to the user in an Email message. The results will list any sequences with the sequence identifiers with similarity to that entered by the user. The complete sequences listed under these identifiers can then be retrieved by the user (using Retrieve or other Email servers) and studied in further detail. The BLAST server is run by the NCBI. Recently, a new program TBLASTX has been available. This will compare a nucleotide query sequence in all six reading frames against a nucleotide sequence database translated in all six reading frames giving a total of 6 x 6 possible combinations. The use of this program is restricted to searching the dbest, dbsts and alu databases on the NCBI BLAST servers. An example of a BLAST search (actual real life query):

Library screening identified a potentially novel cDNA. The cDNA was partially sequenced and yielded the following sequence:

```
tcgacagaaaatgagggaacacctagagttgttttggtcaagagtgaatatcctaag
gtgctgagagccgcagagcaagcccatctttgggcagagctcctctttctgtatgac
aagtatgaagagtatgataagcc
```

The sequence was then submitted to the BLAST Email server with an Email message composed in the following way. No subject need be entered.

```
Program blastn
Datalib nr
Expect 0.75
Begin
>Clone1<
tcgacagaaaatgagggaacacctagagttgttttggtcaagagtgaatatcctaag
gtgctgagagccgcagagcaagcccatctttgggcagagctcctctttctgtatgac
aagtatgaagagtatgataagcc
```

Program blastn	Mandatory. Blastn tells the server to treat the sequence as a nucleotide sequence
Datalib nr	Mandatory. Datalib tells the server which database to search and "nr" (non-redundant) specifies the combined nucleic acids database.
Expect 0.75	The expect parameter is optional. 0.75 specifies a value for expec". This is the signiflcance threshold for reporting matches. The default is 10. Lower threshold matches are more stringent.
Begin	This marks the end of the commands. The next part of the message is to be used for the search. This can be a keyword or an accession number.
>Clone1<	Description line of message.
tcgacaga...	Sequence to use in the search.

Only one database can be searched per Email message. A number of additional directives can be used to control the server and these are listed in the BLAST manual. "blastn" specifies the name of the program and is used for nucleotide sequences. For a protein sequence use "blastp". This is recommended for both protein and nucleic acid database searching as it provides more comprehensive searches with more concise reports. This is because the "nr" database is produced by merging entries from multiple databases into single entries. The "nr" nucleotide database includes sequences from PDB, Genbank(R), GenBank (R) updates, EMBL, and EMBL updates. The message is then followed by the sequence in the FASTA/Pearson format. In the FASTA/Pearson format the sequence begins with a single line description "> clone1". Any text may be used. This is then followed by the sequence in a new line. Each line must be <80 characters in length and only one query sequence per message is permitted. This should be followed by a blank line so that the server does not confuse any signature lines with the sequence itself . The sequence should be in the standard IUB/IUPAC amino acid and nucleic acid codes.

In the example used, the sequence submitted to the BLAST Email server produced a 96% sequence identity with rat clathrin - and the server was also able to align the sequences. In this instance, the sequence was cloned from a human library indicating that the cDNA cloned represented human clathrin. Several other sequences were also identified with less sequence similarity to the query sequence.

Although BLASTN achieves high-speed comparison of nucleotide sequences the user can achieve greater sensitivity by comparing protein sequences with BLASTX, TBLASTN or TBLASTX. This is mostly due to the degeneracy in the genetic code and functional limitations imposed on the protein. Users should submit BLASTN and BLASTP requests at a rate of 1 per 2 minutes and no faster. BLASTX, TBLASTN and TBLASTX requests should be submitted at a rate no faster than 1 per minute. The best time for sending multiple blast jobs is 10 pm - 7 am Eastern Standard time.

The BLITZ Server

The BLITZ server compares the user's protein sequences with the Swiss-Prot database using the Smith and Waterman algorithm (Smith and Waterman, 1981). The program used is the MPsrch program of Sturrock and Collins (1993) and is the fastest implementation of the Smith-Waterman algorithm currently available. The format of the Email message is similar to that used by Mail-FASTA and QUICKSEARCH in that the "Seq" command is mandatory and should precede the sequence. The sequence should then be followed by "end" as a separate line. An example of a message:

```
Title Protein_XYZ A Very Important Protein
Seq
acdefghiklmnpqrs
end
```

Additional commands:

PAM n	This command sets the amino acid weight matrix. 1<n<500. Default is 120.
INDEL n	5<n<30. Default depends on the PAM value chosen.
ALIGN n	Number of best alignments. Maximum is 100. Default is 30
NAMES n	Number of scores to report.
TITLE	Specifies a name for the server to use for the query. In the example above the server will use "protein xyz" as the name and " A very important protein" as the description.

Finding Protein Patterns and Motifs

Assuming the molecular biologist has successfully determined that his/her cloned cDNA sequence is unique and there is a single open reading frame producing a new protein, the next step would be to determine if the protein has any patterns or motifs which would give a hint to the function of the protein (if this is not known yet). Useful servers for motifs and protein patterns:

1.   Motif   `motif @genome.ad.jp`
2.   EBI Mail-Prosite   `prosite@ebi.ac.uk`

Using MotifFinder

MotifFinder will examine the users sequence and then search a Motif library for common patterns (Bairoch, 1992; Ogiwara, 1992). The message is sent to the server in the FASTA format:

```
>protein_VIP  A Very Important Protein
ACDEFGHIKLMNPQRSTVWY
```

>protein...   The subject line of the message. The name of the protein must be preceded by ">".

ACDEFGH...   This is followed by the sequence in single letter amino acid code in a separate line.

If any sequence motifs are found, these will be reported along with the location of the sequence. MotifFinder will also return 3D graphical representations on two separate Postscript files which can be viewed on a Postscript viewer or printed out on a Postscript printer. The default library used is Prosite (Bairoch, 1992) but the user can specifiy the MotifDic (Ogiwara *et al.*, 1992) using the DATALIB command: `DATALIB MotifDic` and this should be inserted before the sequence.

Concluding Remarks

This chapter should provide a starting point for sequence manipulation using facilities available on the Internet. A discussion of all the servers available is beyond the scope of this chapter but the author has attempted to provide a detailed overview of all the servers and information on some of the servers which are likely to satisfy the most of the sequence analysis needs of the molecular biologist.

References

Bairoch, A. 1992. PROSITE: A dictionary of sites and patterns in proteins. Nucleic Acids Res. 20: 2013-2018.

Devereux, J., Haeberli, P. and Smithies O. 1984. A comprehensive set of sequence analysis programs for the VAX. Nucl. Acids Res. 12: 387-395.

Pearson, W.R. and Lipman D.J. 1988. Improved tools for biological sequence comparison. Proc. Natl. Acad. Sci. 85: 2444-2448.

Needleman, S.B. and Wunsch, C.D. 1970. A general method applicable to the search for similarities in the amino acid sequence of two proteins. J. Mol. Biol. 48: 443-453.

Ogiwara, A., Uchiyama, I., Seto, Y. and Kanehisa, M. 1992. Construction of a dictionary of sequence motifs that characterize groups of related proteins. Protein Eng. 5: 479-488.

Smith, T.F. and Waterman, M.S. 1981. Identification of common molecular subsequences. J. Mol. Biol. 147: 195-197

Sturrock, S.S. and Collins, J.F. 1993. MPsrch version 1.3. Biocomputing Research Unit, University of Edinburgh, UK.

Further Reading

Altschul, S.F., Gish, W., Miller, W., Myers, E.W. and Lipman D.J. 1990. Basic local alignment search tool. J. Mol. Biol. 215: 403-410.

Bairoch, A. 1995. List of molecular biology Email Servers. (The author has made extensive use of this list). Amos Bairoch's Email address is bairoch@cmu.unige.ch. The latest version is available is available by FTP on the server expasy.hcuge.ch (129.195.254.61) and is called serv_ema.txt in the directory: /databases/info.

Henikoff, S. 1993. Sequence analysis by electronic mail server. Trends in Biochem. Sci. 18: 267-268.

From: *Internet for the Molecular Biologist.*
ISBN 1-898486-02-6 ©1996 Horizon Scientific Press, Wymondham, U.K.

8

COMPUTATIONAL GENE IDENTIFICATION

James W. Fickett and Roderic Guigó

Introduction

In this chapter, we present a guide to Internet resources for identifying genes in nucleic acid sequences. By "identifying" we mean both locating the genes and, when possible, assigning them a tentative function. There has been a great deal of progress in gene identification methods in the last few years. At least in the case of sequence data from mammals, *C. elegans*, and *E. coli*, the older coding region identification methods have mostly given way to methods that can suggest the overall structure of genes. For all organisms, computational methods are sufficiently accurate that they give practical help in many projects of biological and medical import.

However the situation with respect to services is not simple. The choice of a program or programs to use depends, for example, on what organism is being studied, whether one is analyzing single sequence runs or large assembled sequences, and how much effort the user is willing to expend. It would be very convenient if one program, in one place, could do everything needed in the way of gene identification, but we are unlikely to enjoy this ideal soon. To help the user choose the most appropriate services for each situation, we will briefly describe the kinds of information that each program or database is capable of supplying. The overall flow of the chapter is intended to provide a protocol for gene identification using a number of services (not all of which need to be included in each case).

Only a small fraction of the relevant background, literature, and resources available can be mentioned here. For more details and other points of view there are a number of related reviews, of which the following are perhaps closest to the goals of the current work: (Gribskov and Devereux 1991; Adams *et al.*, 1994; Gelfand 1995; Snyder and Stormo 1995b). Most programs require a particular sequence format, have limits on the length and number of sequences that may be submitted for analysis and, except in the case of database searches, are designed for only one or a few specific organisms. These details are constantly changing and will not be given here. Instead we will describe how to obtain the current documentation for the service. Only one recent reference is given for each service; usually it will contain references to related work.

Emphasis is given to services, that is, to programs that are available for remote execution over the Internet. All these services respect the privacy of their users, and guarantee not to keep any record of sequences analyzed. However some investigators may, for maximum security or for higher throughput, prefer to install the analysis programs locally. Thus we also mention, where applicable, availability of source code. Similarly, we will also mention programs which are not strictly Internet servers, but whose source code or executables can be obtained through the Internet. Network access information is given for each service. Unless otherwise mentioned, an address labeled "Email" is for an Email server program; one labeled "inquiries" is answered personally by one of the developers of a program; "FTP" is for an anonymous FTP site where source code or data may be retrieved, and "WWW" is for a World Wide Web page with information about the program and, sometimes, interactive use of the program.

Finding and Masking Repeats

In eukaryotic sequences, repeats should be masked early in the process to:

* Help delimit where coding regions may occur
* Eliminate meaningless, voluminous output from database searches
* Avoid confusing the gene identification programs.

Pythia

Reference   Milosavljevic 1995
Email   send message "help" to pythia@anl.gov

Using Pythia

Pythia removes the repeats and returns the non-repetitive parts of the sequence separately. When the following sequence (in Stanford format) is sent to pythia@anl.gov, in an Email message with subject "Rpts",

```
;HUMCKMM1 (GenBank) Human muscle creatine kinase gene (CKMM)
HUMCKMM1
GGATCCTTCCTCCTTGGCCTCCCAAAGTGCTGGGATTACAGGTGTGAGCCACTGCACCTGGCCTATTACCCTTCTCAGGCTCTCTGGAGTCC
ATCCTTCTGCTCTGTCTCCCTCAGTTCAATTGTTTTTTGTTTTTTTGTTTTTTTTTTTAGACACAGTCTCGCTCTGTCACCAAGGCTGGAGT
GCAGCAGTGCGATCACAGCTCACCGCAGCCTCACCTCCCAGGCTCAAGTGATCCTCCCATCTCGGCCTCTGAGTAGCTGAGACTATAGGT
GTGTCCACATGTCCGGCTAATTTTTGTATTTTTAGTAGAGACAGGGTTTCACCGCGGTTGGCCAGGGTGGTCTTGAACTCCTGAGCTCAAG
CAATCCTCCTGCCTCAGCCTCCTTGTTTTGATTTTTAGATCCCACAAATAACTTGTGATGTTTGTCTTTCTATACCTGGTTCATTTAACA
TTTTCTTTTTCTTTTCTTTTCTTTTTTTTTTTTTTTTTGTGAGACTGAGTCTTGCTCTGTCACTCAGGCTGGAGGGCAATGGTGCATCTCAG
CTCACTGCAACCTCCACCTCCTAGGTTCAAGCAATTCTTATGCCTCAGCCTCCTGGCTAGCTGGGATTACAGGCGTGTGTCACCATGCCA
GGCTAATTTTTGTACTTTTAGTAGAGATGGGGTTTCACCATGTTGGCCAGGCTGGTCTTGAACTCCTGGCCTCAAGTGATCCACCCGGCCT
CCGGCCTCTGCCTCCCAAAGTGCTGGGATTACGGGCCTGAGCCACTGTGCCCGGCCCATCTAACATTTTCACTGTCAATCACAATGGGATT
AAAACTCCTCCCACAGCCCCTAGGGACCA1
```

the result is an extensive and informative analysis of the repeat structure of the sequence, and the following masking results.

```
(4) Listing of non-repeat fragments:
------------------------------------------------------------------------

;HUMCKMM1(1->1)(0,0)
G1
;
HUMCKMM1(64->137)(0,0)
TATTACCCTTCTCAGGCTCTGGAGTCCATCCTTCTGCTCTGTCTCCCTCAGTTCAATTGTTTTTTGTTTTTTGT1
;
HUMCKMM1(383->480)(0,0)
TTGTTTGATTTTTAGATCCCACAAATAACTTGTGATGTTTGTCTTTCTATACCTGGTTCATTAACATTTTCTTTTTCTTTTCTTTTCTTTTTTTT1
------------------------------------------------------------------------
```

xblast

Reference	Claverie 1992 and 1994a
Inquiries	`jmc@euler.cnrs-mrs.fr`
FTP	`ncbi.nlm.nih.gov` directories `pub/jmc/xblast`
	and `pub/jmc/alu`

Using xblast

xblast leaves the sequence intact, overlaying repeats with a character of the user's choice. xblast requires the user to set up the software locally, but once this is done it runs very quickly. The same sequence as above, processed by xblast, gives the following results.

```
>HUMCKMM1 (GenBank) Human muscle creatine kinase gene (CKMM)
Gnnnnnnnnnnnnnnnnnnnnnnnnnnnnnnnnnnnnnnnnnnnnnnnnnnnnnnnnnnnnnn
nnnTATTACCCTTCTCAGGCTCTGGAGTCCATCCTTCTGCTCTGTCTCCCTCAGTTCAAn
nnnnnnnnnnnnnnnnnnnnnnnnnnnnnnnnnnnnnnnnnnnnnnnnnnnnnnnnnnnnnn
nnnnnnnnnnnnnnnnnnnnnnnnnnnnnnnnnnnnnnnnnnnnnnnnnnnnnnnnnnnnnn
nnnnnnnnnnnnnnnnnnnnnnnnnnnnnnnnnnnnnnnnnnnnnnnnnnnnnnnnnnnnnn
nnnnnnnnnnnnnnnnnnnnnnnnnnnnnnnnnnnnnnnnnnnnnnnnnnnnnnnnnnnnnn
nnnnnnnnnnnnnnnnnnnnnnnnnnnnnnnnnTGATTTTTAGATCCCACAAATAACTTGTGATG
TTTGTCTTTCTATACCTGGTTCATnnnnnnnnnnnnnnnnnnnnnnnnnnnnnnnnnnnnn
nnnnnnnnnnnnnnnnnnnnnnnnnnnnnnnnnnnnnnnnnnnnnnnnnnnnnnnnnnnnnn
nnnnnnnnnnnnnnnnnnnnnnnnnnnnnnnnnnnnnnnnnnnnnnnnnnnnnnnnnnnnnn
nnnnnnnnnnnnnnnnnnnnnnnnnnnnnnnnnnnnnnnnnnnnnnnnnnnnnnnnnnnnnn
nnnnnnnnnnnnnnnnnnnnnnnnnnnnnnnnnnnnnnnnnnnnnnnnnnnnnnnnnnnnnn
nnnnnnnnnnnnnnnnnnnnnnnnnnnnnnnnnnnnnnnnnnnnnnnnnnnnnnnnnnCATCT
AACATTTTCACTGTCAATCACAATGGGATTAAAACTCCTCCCACAGCCCCTAGGGACCAT
```

Database Searches

To identify possible homologues, compare the sequence to both the nucleotide and protein sequence databases (this section). Then check the sequence for occurrence of functional motifs, by means of the BLOCKS, or ProSite/MotifDic collections (next section). It seems that currently about half of all newly sequenced genes can be expected to have a detectable homologue in the sequence and motif databases (Green *et al.*,

1993; Koonin *et al.*, 1994). This situation may change with the very rapid accumulation of publicly available cDNA sequences see Williamson *et al.*, (1995) and `http://genome.wustl.edu/est/esthmpg.html`.

For nucleotide and protein sequences, the databases may be searched either by downloading the data and search program, and searching locally, or by using remote, on-line services. Essentially the same nucleotide sequence data is available from any of the four major databases: DDBJ (DNA DataBank of Japan), the European Bioinformatics Institute (EBI) data library, GenBank, and GSDB (Genomes Sequence DataBase). Public cDNA sequences are maintained both in the main collection and, with more detailed information, in dbEST.

Databases

DDBJ	Inquiries	`ddbj@ddbj.nig.ac.jp`
	FTP	`ftp.nig.ac.jp`
EBI	Inquiries	`datalib@ebi.ac.uk`
	FTP	`ftp.ebi.ac.uk`
	WWW	`http://www.ebi.ac.uk`
GenBank	Inquiries	`info@ncbi.nlm.nih.gov`
	FTP	`ncbi.nlm.nih.gov`
	WWW	`http://www.ncbi.nlm.nih.gov`
GSDB	Inquiries	`gsdb@gsdb.ncgr.org`
	WWW	`http://www.ncgr.org`
dbEST	WWW	`http://www.ncbi.nlm.nih.gov/dbEST/index.html`

For protein sequences, the Swiss-Prot collection is well known for being carefully curated and highly reliable. On the other hand, because such curation takes time, one may sometimes find sequences in other collections that have not yet found their way into Swiss-Prot. Probably the most complete collection is the "non-redundant" protein sequence database compiled at NCBI (National Center for Biotechnology Information) from PDB, Swiss-Prot, PIR(R), and translated coding regions from GenBank, with duplicates removed. The non-redundant database is available for searching via the BLAST program described below.

Swiss-Prot	FTP	`expasy.hcuge.ch`
	WWW	`http://expasy.hcuge.ch/`

Database Search Programs

One of the most commonly used databases search servers is that for BLAST (Altschul *et al.*, 1990) at NCBI. The BLAST Email server has many convenient options, such as automatically translating a nucleotide sequence in all six frames and using the translations to search either of the protein databases mentioned above. The software is also available. A European service may be found in the FASTA (Pearson and Lipman, 1988) server at EBI. The software is available from the author. Both BLAST and FASTA achieve very high speed searches at the cost of possibly (though with very low probability) missing some statistically significant matches. There are some servers that use special purpose hardware to carry out an exhaustive search with an acceptable turnaround time. These include bicserv, BLITZ (Sturrock and Collins, 1993), and dFLASH (Califano and Rigoutsos, 1993).

BLAST	Inquiries	`blast-help@ncbi.nlm.nih.gov`
	Email	send message "help" to `blast@ncbi.nlm.nih.gov`
	FTP	`ncbi.nlm.nih.gov`, directory `blast`
FASTA	Inquiries	W.R. Pearson at `wrp@virginia.edu`
	Email	send message "help" to `fasta@ebi.ac.uk`
	FTP	`ftp.virginia.edu` directory `pub/fasta`
bicserv	Email	send message "help" to `bicserv@sgbcd.weizmann.ac.il`
BLITZ	Email	send message "help" to `blitz.ebi.ac.uk`
dFLASH	Email	`dflash@watson.ibm.com` send message "help", subject "dFLASH".

To illustrate the value of the nucleotide-level search, we give a sample query using an enhancer region. The upstream enhancer of the mouse MCK gene is formatted as follows to send to BLAST (`blast@ncbi.nlm.nih.gov`)

```
PROGRAM blastn
DATALIB nr
BEGIN
>mouse mck upstream enhancer
ctatgggtctaggctgcccatgtaaggaggcaaggcctggggacacccgagatgcctggttataattaacccagacatgtggctgctccc
cccccccaacacctgctgcctgagcctcacccccacccccggtgcctgggtcttaggctctgtacaccatggaggagaagctcgctctaaa
aataaccctgtccctggtgga
```

In the sample output below, the core information in the response from BLAST is a list of the high-scoring hits. The next-to-last column is an estimate of the probability that the match could happen by chance. It will be seen that the first four hits are estimated to be significant. In this case it is also clear that they are homologues.

Sequences producing High-scoring Segment Pairs:		High Score	Smallest Sum Probability P(N)	N
gb|M21390|MUSMCKA	Mouse muscle creatine kinase (MCK) g...	1005	5.2e-76	1
gb|M27092|RATCKMUSCL	Rattus norvegicus muscle creatine ki...	357	1.6e-45	3
gb|M21487|HUMCKMM1	Human muscle creatine kinase gene (C...	188	5.9e-23	4
emb|X55146|OCMCK1	Rabbit gene for muscle creatine kina...	93	0.0090	3
emb|X54020|HSPPAE5	Human preproacrosin gene exon 5 (EC ...	121	0.92	1
gb|M77381|HUMACRO4	Human acrosin gene, 3' end.	121	0.92	1
emb|Y00970|HSACROS	Human mRNA for acrosin (EC 3.4.21.10)	121	0.93	1
gb|R11426|R11426	yf46d06.r1 Homo sapiens cDNA clone 1...	116	0.998	1
emb|X54053|MMKFGF5	Mouse k-FGF oncogene 5' sequence.	94	0.999	2
gb|J05595|RICLNOCI	Rice oryzacystatin-II mRNA, complete...	116	0.999	1
emb|X57658|OSORYII	O.sativa oryzacystatin-II gene	116	0.9992	1

Further down in the response mail from BLAST, the alignments themselves are given, for example

```
>emb|X55146|OCMCK1 Rabbit gene for muscle creatine kinase, exon 1
```

```
Length = 804

Plus Strand HSPs:

Score = 93 (25.7 bits), Expect = 0.0091, Sum P(3) = 0.0090
Identities = 21/24 (87%),Positives = 21/24 (87%),Strand = Plus / Plus

Query:     18 CCATGTAAGGAGGCAAGGCCTGGG 41
              ||  |||||||||||||||||| ||
Sbjct:    145 CCTTGTAAGGAGGCAAGGCCCAGG 168


Score = 87 (24.0 bits), Expect = 0.0091, Sum P(3) = 0.0090
Identities = 23/30 (76%),Positives = 23/30 (76%),Strand = Plus / Plus

Query:    168 AGCTCGCTCTAAAAATAACCCTGTCCCTGG 197
              ||  |||||||||||| | |  |||||
Sbjct:    276 AGACCGCTCTAAAAATAGCTCCCTTCCTGG 305


Score = 73 (20.2 bits), Expect = 0.0091, Sum P(3) = 0.0090
Identities = 17/20 (85%), Positives = 17/20 (85%), Strand = Plus / Plus

Query:     41 GGACACCCGAGATGCCTGGT 60
              |||||||||||| ||||| |
Sbjct:    167 GGACACCCGAGCTGCCGGCT 186
```

For a BLAST comparison of a nucleotide sequence, conceptually translated in all six reading frames, with the protein sequence databases, the following format is used.

```
PROGRAM blastx
DATALIB nr
BEGIN
>human desmin exon
1ggtaggcttgccgtcacaggacccccgctggctgactcaggggcgcaggctcttgcggggggagctggcctcccgccccacggccacgg
gccctttcctggcaggacagcgggatcttgcagctgtcaggggaggggaggcgggggctgatgtcaggagggatacaaatagtgccgacg
gctgggggcctgtctccctcgccgcatccactctccggccggccgcctgcccgccgcgctcctccgtgcgccgccagcctcgcccgcg
ccgtcaccatgagccaggctactcgtccagccagcgcgtgtcctcctacacgccgcacctcggcgcggcgcccccggtcttctcgctctggct
ccccgctgagctcgcccgtgttcccgcgggcgagccttttcggctctcaaggcctcctccagctcggtgacgtcccgcgtgtaccaggtgtcgc
gcacgtcgggagggcccgggggcctgggctcgctgcgggggcagccggctgggaccacccgcacgccctcctcctacggcgcaggcgagc
tgctggacttcttcactggccgacgcggtgaaccaggagtttctgaccacgcgcaccaacgagaaggtggagctgcaggagctcaatgacc
gttcgccaatctacatggagaaggtgcgcttcctggagcagcagaacgcgctcgccgccgaagtgaacccggctcaagggccggagccga
cgcgagtggccgagctctacgaggaggagctgctgcgggagctgcggcgccaggtggaggtgctcactaaccagcgcgcgcgggtcgacgtcg
agcgcgacaacctgctcgacgacctgcagcgggctcaaggccaagtgaggggcccggcacccccagactcctcttttctgcgggcagggcacag
gaggctaggcctgggggctggggtcccgctgtcaccacctgccttctcccgggggcccgacgctcctccccatgtgtggagaaagggtcct
ccacctgtgtgtgtttcaaggggccgtgacctc
```

The value of such a search is shown clearly by the response, which includes "hits" on several homologous proteins:

		Reading Frame	High Score	Smallest Sum Probability P(N)	N
pir|JE0063|DMHU	desmin - human >gp|M63391|HUMDE...	+2	947	1.1e-126	1
sp|P17661|DESM_HUMAN	DESMIN.	+2	942	5.5e-126	1
gp|X73524|RNDES_1	desmin (Rattus norvegicus)	+2	377	5.4e-119	3
pir|A24783|A24783	desmin - hamster >gp|M12104|HAM...	+2	374	1.4e-118	3
pir|A54104|A54104	desmin - mouse	+2	371	2.6e-118	3
sp|P02541|DESM_MESAU	DESMIN.	+2	369	6.8e-118	3
sp|P02542|DESM_CHICK	DESMIN. >pir|A90969|DMCH desmin...	+2	238	5.0e-85	4

```
sp|P23239|DESM_XENLA   DESMIN. >pir|A43554|A43554 desm...  +2   245  2.2e-61  3
sp|P20152|VIME_MOUSE   VIMENTIN.                           +2   189  4.9e-41  3
sp|P31000|VIME_RAT     VIMENTIN.                           +2   189  4.9e-41  3
...
sp|P41219|PERI_HUMAN   PERIPHERIN. >pir|A55185|A55185 ...  +2   179  3.4e-35  3
...
gp|Z48978|RNGFABEX1_1  glial fibrillary acidic protein...  +2   173  9.5e-34  2
...
sp|P23729|IF3T_TORCA   TYPE III INTERMEDIATE FILAMENT....  +2   188  5.3e-32  2
...
```

Searching for Functional Motifs

Searches for functional motifs in protein sequences may be carried out against the BLOCKS database (Henikoff and Henikoff, 1994), or the PROSITE (Bairoch, 1992) and MotifDic (Ogiwara *et al.*, 1992) collections. The former service, illustrated below, automatically detects a nucleotide sequence, translates it in all six frames, and compares the resulting amino acid sequences against the motif database. The latter service (send "help" to motif@genome.ad.jp) requires an amino acid sequence as input.

BLOCKS

Inquiries   henikoff@howard.fhcrc.org
Email   send message "help" to blocks@howard.fhcrc.org

Using BLOCKS

Sending the following query, containing a desmin gene, to the BLOCKS service (100 lines of sequence have been left out of the display for brevity).

```
>HUMDES (GenBank) Human desmin gene, complete cds.
CCTCCGTGCGCCCGCCAGCCTCGCCCGCGCCGTCACCATGAGCCAGGCCTACTCGTCCAG
CCAGCGCGTGTCCTCCTACCGCCGCACCTTCGGCGGCGCCCCGGTCTTCTCGCTCGGCTC
...
CACCTGGCCCTGAAATCTTAAAGGGAGATAGGTACTGTAAGGTCCTCTAAAGAGTGTCTT
```

results in a strong match to a set of blocks characteristic of intermediate filament proteins:

```
Query=HUMDES (GenBank) Human desmin gene, complete cds. ,
  Size=6780 Base Pairs
Database=mats.dat, Blocks Searched=2884
```

```
1.---------------------------------------------------------------------
Block     Rank Frame Score Strength   Location (bp) Description
BL00226A    2    2   1519  1539          356-   434 Intermediate filaments p
BL00226B    3    1   1371  1460         1816-  1933 Intermediate filaments p
BL00226C    1    1   1586  1549         3004-  3136 Intermediate filaments p
```

```
1586=100.00th percentile of anchor block scores for shuffled queries
P not calculated for single block BL00226C
                        |- 167 amino acids-|
   BL00226 AAAA::::..............BBBBBB:::::::::::::::::::.......CCCCCCC
   HUMDES <:::::::::::::::::::::::::::::::::::::::::::::::::::::::::CCCCCCC
```

Each of the three blocks in the intermediate filament protein motif give a strong match to the submitted sequence. Introns in the genomic sequence prevent the software from detecting that the three blocks are correctly spaced in the protein product of the gene.

PROSITE/MotifDic

Email   Send message "help" to `motif@genome.ad.jp`
Inquiries   `motif-manager@genome.ad.jp`

Gene Identification

In this section we describe services that emphasize finding whole genes, or at least sets of exons. These services are usually highly organism dependent (the documentation will give the current range of organisms supported), and only give reasonable results if the sequence is long enough to contain one or more full exons. The next section describes services offering coding region analysis, which is better for short sequences, offers higher throughput, and is usually not quite as dependent on the organism.

Integrated gene identification algorithms, which combine a coding region analysis with recognition of ribonucleotide or protein binding sites, all in the context of a simple overall 'grammar', are still fairly new. The accuracies reported by developers are almost impossible to compare, due to different testing methodology. Some third-party benchmarking is available (Singh and Krawetz, 1993; Lopez *et al.*, 1994; Snyder and Stormo, 1995b; M. Burset and R. Guigó unpublished data), which we may summarize as follows:

- On sequences containing a single, entire gene, from an organism for which the algorithm is specifically intended, current programs will predict a gene (or exons) having significant overlap (55-95%) with the known one, but usually having significant differences as well.
- If the sequence contains multiple, partial, or alternatively spliced genes, the performance of most algorithms is currently unknown.
- Performance on genes not homologous to those studied to date is noticeably worse than average.
- Performance seems to vary widely between algorithms (the biggest differences seem to depend on whether database search results are directly incorporated in the algorithm), and significant differences in accuracy are given when different groups test the same algorithm on different data sets.
- Performance for most programs is quite sensitive to sequence errors.

Since different algorithms have complementary strengths, we strongly recommend first sending a few sequences with known gene structure to several of the services, to get a feeling for algorithm capabilities. Then the sequence under analysis should be sent to the same services and the results compared.

BCM Genefinder System

Reference   Solovyev *et al.*, 1994
Email   send message "help" to `service@bchs.uh.edu`
WWW   `http://condor.bcm.tmc.edu/Genefinder/`
            `genefinder.html`

Description

Several services for the analysis of human DNA sequences are offered: prediction of splice sites (HSPL), prediction of internal exons (HEXON) and 5'- and 3'- exons (FEXH), and prediction of gene structure (FGENEH). HSPL uses a linear discriminant function to score AG and GT dinucleotides as potential acceptor and donor sites. The value of the function is computed from the value of several individual measures related to oligonucleotide composition upstream and downstream of the site. HEXON predicts internal exons by using a linear discriminant function to score ORFs flanked by AG and GT dinucleotides. The individual variables for which the function is computed include octanucleotide preferences upstream, downstream and within the ORF (Open Reading Frame), and the splice site recognition function value (as computed by HSPL). FEXH predicts internal exons exactly as HEXON, but in addition, it predicts initial and terminal exons. Initial and terminal exon recognition modules similar to those developed for internal exons are used to scan the first predicted internal exon, and score potential initial exons (ORFs flanked by ATG and GT), and to scan the last predicted internal exon and score potential terminal exons (ORFs flanked by AG and stop codon). FGENEH finds all possible internal exons as in HEXON, and potential initial and terminal exons within each predicted internal exon. It then uses dynamic programming to find a combination of exons with best overall score, to construct a gene model. (Dynamic programming is a well known optimization technique that can be shown to always find the top-scoring solution. Of course this does not guarantee that the solution is biologically correct).

        Each sequence requires a separate Email message. The service requested (FGENEH, FEXH, HEXON, or HSPL) must be provided in the subject line. The first line of the message must be the name of the sequence. The following lines contain the sequence. If the service requested is HSPL, an additional line must be placed after the sequence name with two numbers, indicating donor and acceptor threshold scores. Both the BCM and GRAIL (see below) services usually respond almost immediately.

Using BCM

A typical message for the BCM Genefinder Email server, requesting an FGENEH service, is shown below:

```
To: service@theory.bchs.uh.edu
Subject: fgeneh
```

> Human enolase 3
```
AGATCTCTACCGAGGGCAGAGACCTACCTCCCCGCAGTGCTACAAGTGGGGCGCCGGAAG
AGCCCCAGGCGTGCAGAAGCTCACAAAAGGCCACCCGTCCTCGGTCCATTCATTTTTGTT
CACTGTTGATTCAGCCCCATTCATTGATGGGCTGGGGCCGTGCGCTGAGCGCCCACAGTC
...
```

The results were as follows:

```
> Human enolase 3
 length of sequence -     7140
 number of predicted exons - 11
 positions of predicted exons:
    354 -      378 w=   10.54
   1577 -     1663 w=    4.76
   2540 -     2635 w=   10.69
   2796 -     2858 w=    5.91
   3455 -     3588 w=    5.59
   4820 -     5042 w=   12.10
   5153 -     5350 w=   12.43
   5688 -     5889 w=    9.95
   6318 -     6426 w=    4.81
   6576 -     6634 w=    8.45
   6723 -     6792 w=    6.42
 Length of Coding region-     1266bp        Amino acid sequence -
421aa
MAVMRTLRAMAMQKIFAREILDSRGNPTVEVDLHTAKGRFRAAVPSGASTGIYEALELRD
GDKGRYLGKGVLKAVENINNTLGPALLQKATKFGANAILGVSLAVCKAGAAEKGVPLYRH
IADLAGNPDLILPVPAFNVINGGSHAGNKLAMQEFMILPVGASSFKEAMRIGAEVYHHLK
GVIKAKYGKDATNVGDEGGFAPNILENNEALELLKTAIQAAGYPDKVVIGMDVAASEFYR
NGKYDLDFKSPDDPARHITGEKLGELYKSFIKNYPVVSIEDPFDQDDWATWTSFLSGVNI
QIVGDDLTVTNPKRIAQAVEKKACNCLLLKVNQIGSVTESIQACKLAQSNGWGVMVSHRS
GETEDTFIADLVVGLCTGQIKTGAPCRSERLAKYNQLMRIEEALGDKAIFAGRKFRNPKA
K*
```

There is also a WWW server available for forms-based WWW clients. The input is provided by cut and paste of the sequence in the fill-out form with the same format required by the Email server. Output is returned as an Email message.

EcoParse

Reference   Krogh *et al.*, 1994
Email   send message "help" to ecoparse@cse.ucsc.edu

Description

EcoParse finds protein coding genes in *E. coli* DNA sequences. EcoParse uses a Hidden Markov Model (HMM) architecture to integrate gene components into overall gene models. An HMM encodes a probabilistic grammar of genes; for each gene component, the model specifies a probability (based on a set of real genes) of passing to any other component. For example, from a start codon must be followed by another codon, and the probability of passing to any particular codon is based on the frequency with which that codon occurs in actual sequences. The HMM of EcoParse models not only codons, gene starts, and gene ends, but also patterns found in intergenic regions. For any syntactically correct parse of the sequence, the HMM assigns on overall probability. Given a query sequence, a standard dynamic programming procedure for HMMs (the Viterbi algorithm) is applied to find the most likely parse. Among the distinctive features of EcoParse are the possibility of predicting overlapping genes, and of accounting for potential sequencing errors and/or frameshifts in the raw genomic DNA sequence.

Using EcoParse

Any number of sequences can be included in a single mail message to EcoParse (ecoparse@cse.ucsc.edu). The server allows a considerable amount of freedom in the sequences format, and even a mixture of formats can be used within the same message. Below is an example (taken from EcoParse help file):

```
tccgttaaactgggaacatcatcagccctgtgtgcgtggcgaacgtggcaatggcaaggattgcgtttcagcgcgcactggcctttacaa
ggtagcaacgataaaggaaataaccattcctgtgtggttaaggttgatgacgggacgaatagcattcttctaaccggtgatattgaagcc
ccagctgaacaaaagatgc

>seq1 Any comments etc can follow the id.
GCAACATTGCTTCAGGTACCTCACCATGGCAGTAATACCTCATCATCGTTGCCATTAATTCAGCGAGTGAATGGA
AAAGTGGCACTCGCATCGGCATCGCGCTATAACGCATGGCGACTGCCCTCTAACAAAGTTAAGCATCGCTATCAA
CTGCAAGGATATCAAT1
                   ATTAGCAGCC TCAGGGAGCA AATTTTACCT CGTTGGTATC ATCAGTGGTT   50
                   GTGGATAACG AATATGCGGC TATTTCAACA AATGCTGGTT TTTTGAATGC 100
                   AGATCTCTCT ACGT 114
>seq2
CGCCGACTGTGGCCAACCATTGCGCCTTTCAAAGCGGGTCTGATCGTGGCGGGCGTAGCGTTAATCCTCAACGCA
GCCAGCGATACCTTCATGTTATCGCTCCTTAAGCCACTTCTTGATGATGGCTTTGGTAAAACAGATCGCTCCGTG
```

A number of options, mostly controlling the format of the output, can be specified. EcoParse's output is the list of predicted genes, one on each line, before and after post-processing. The first two columns indicate the position of the gene, and the third column is the gene index—the negative log probability of the gene under the HMM—. Additional columns may follow, indicating predicted errors or frameshifts. A display of the sequence with the discovered genes is also returned. Below is a (trimmed) example of EcoParse output:

```
----------------LIST OF PREDICTIONS------------------------------------------------

 *** AFTER  post-processing

TestSeq 3960
            3531   3960   0.917   3959 -2

----------------LIST OF PREDICTIONS-------------------------------------------------

*** BEFORE  post-processing

TestSeq 3960
            3531   3960   0.917   3959 -2
.
---------DISPLAY OF PREDICTIONS---------------------------------------------------

 *** AFTER  post-processing
```

```
TestSeq

ACTCAGCCCCAGCGGAGGTGAAGGACGTCCTTCCCCAGGAGCCGGTGAGAAGCGCAGTCGGGGGCACGGGGA    72
.........................................................................
.........................................................................
GGGAATAGAGACATGAGCCACCTTGCTCGGCCTCCTAGCTCTTTCTCCGTCTCTGCCTCTGCTCTCTGCGTC   3312
TGTCTTTGTCTCCTCTCTGCCTCTGTCCCGTTCCTTCTCTCTTGGTCCACTGCCCTTCTGTCTCTCCCTGTT   3384
CTCCTTAGGAGACTCTCCTCTCTTCCTTCTCGAGTCTCTCTGGCTGATCCCCATCTCACCCACACCTATCCC   3456
AGCCCTTCTCCTCGCCTCCCCCTGTGCGCACACCCTCCCGCTCTTTCGGCTGCAGGACGCTGATGGACGAGA   3528
CCATGAAGGAGTTGAAGGCCTACAAATCGGAACTGGAGGAACAGCTGAGCCCGGTGGCGGAGGAGACGCGGG   3600
>>>***>>>***>>>***>>>***>>>***>>>***>>>***>>>***>>>***>>>***>>>***>>>*

CACGGCTGTCCAAGGAGCTGCAGGCGGCGCAGGCCCGGCTGGGTGCCGACATGGAGGACGTGCGCAGCCGCC   3672
**>>>***>>>***>>>***>>>***>>>***>>>***>>>***>>>***>>>***>>>***>>>***>>>*

TGGTGCAGTACCGCAGCGAGGTGCAGGCCATGCTGGGCCAGAGTACCGAGGAGCTGCGGGCGCGCCTCGCCT   3744
**>>>***>>>***>>>***>>>***>>>***>>>***>>>***>>>***>>>***>>>***>>>***>>>*

CCCACCTGCGCAAGCTGCGCAAGCGGCTCCTCCGCGATGCTGATGACCTGCAGAAGCGCCTGGCAGTGTATC   3816
**>>>***>>>***>>>***>>>***>>>***>>>***>>>***>>>***>>>***>>>***>>>***>>>*

AGGCCGGGGCCCGCGAGGGCGCCGAGCGCGGGGTCAGCGCCATCCGCGAGCGCCTGGGACCCCTGGTGGAGC   3888
**>>>***>>>***>>>***>>>***>>>***>>>***>>>***>>>***>>>***>>>***>>>***>>>*

AGGGCCGCGTGCGGGCCGCCACTGTGGGCTCCCTGGCCAGCCAGCCGCTTCAGGAGCGGGCCCAGGCCTTGG   3960
**>>>***>>>***>>>***>>>***>>>***>>>***>>>***>>>***>>>***>>>***>>>***>>>*
```

GeneID

Reference	Guigó *et al.*, 1992
Email	send message "help" to `geneid@bir.cedb.uwf.edu`
WWW	`http://www.imim.es/GeneIdentification/Geneid/geneid_input.html`

Description

The GeneID service offers, for vertebrate sequences, both prediction of genic structure in genomic sequences and coding region location in cDNA sequences. GeneID is hierarchical. For genomic sequences (cDNAs are similar) this means that the program begins by identifying and scoring potential start and stop codons and splice sites, builds up potential initial, internal and terminal exons from these, and then builds gene models from the exon pool. At each stage, several tests are applied, and built in rules are used to select the promising structures. For example, at the exon level, a number of coding statistics are calculated and those exons are discarded for which any of the coding measures indicates low coding likelihood. Remaining exons are scored by a pre-trained neural network whose input values are the scores of the exon defining atomic sites and the exon coding measures. (Neural networks are one of the more common methods for weighting and combining several pieces of evidence. See (Lippmann, 1987) for an introduction.) Next the translated sequence is used to query the SWISSPROT database, and exons overlapping any regions showing similarity to known proteins are rescored with a combination of the extent of overlap, the strength of the similarity and the original GeneID score. At the gene assembly level, exons are clustered into equivalence classes, with two exons being considered equivalent if they can occur in exactly the same gene models. Only the highest scoring exon within each class is further considered. Gene models are scored as function of the constituent exon scores. An exhaustive search of the space of gene models is performed. GeneID's result is a set of candidate gene models ranked according to score. A particular advantage of GeneID is the inclusion of database search results in the evaluation of potential exons. GeneID can also be used to predict potential PolII promoters in genomic sequences (Knudsen, S., unpublished), but this prediction is not yet incorporated into the identification and assembly of genes.

Using GeneID

Only one sequence can be sent per Email message to the server (geneid@bir.cedb.uwf.edu). The sequence, in FastA format, is placed after a line containing either "Genomic Sequence" or "cDNA", as shown in the example below.

```
To: geneid@bir.cedb.uwf.edu
Subject:

Genomic Sequence
> Human enolase 3
AGATCTCTACCGAGGGCAGAGACCTACCTCCCCGCAGTGCTACAAGTGGGGCGCCGGAAG
AGCCCCAGGCGTGCAGAAGCTCACAAAAGGCCACCCGTCCTCGGTCCATTCATTTTTGTT
CACTGTTGATTCAGCCCCATTCATTGATGGGCTGGGGCCGTGCGCTGAGCGCCCACAGTC
...
```

Different levels of output can be specified. GeneID's most exhaustive output reports potential Pol III promoters, atomic sites, potential exons before and after classification at different levels of confidence, exon matches to amino acid sequences, if any, and predicted gene models. In addition, it can report the result of comparing the translation of the top ranking predicted gene model to the protein sequence databases, using the BLAST Network Service provided by NCBI. A (trimmed) example of GeneID's output follows.

```
REPORT for genomic DNA sequence Human enolase 3, 7194 bases.
Fri Jul 28 10:20:56 CDT 1995


0. Potential PolII PROMOTERS
Promoter prediction:
  Region      Score  Likelihood
    0 -  200  0.608  Marginal prediction
  700 - 1000  1.140  Highly likely prediction
...


1. potential START CODONS

    Position  Score  Sequence
       147    0.70   ATTGATGG
       182    0.76   GTCGATGG
...
```

```
5a. potential FIRST EXONS (from startcodon to donor site)
      Score  Frame  Start  Stop           Blastx matches    Blast score

      —Highly confident predictions, more than 50% true exons: —
      1.66   1        4919  5042                            1.37
      1.64   1        1579  1663                            1.49
      1.53   1        4868  5042        sp|P25704|ENOB_RABIT  1.22
...


8. potential GENES


55611 gene models were analyzed and ranked according to likelihood
The 20 most likely genes are

Ranking scores   List of exons (*) constituting gene (first, internal(s),
last)
  1.969 11.227      354    378      1577  1663      2540  2635      3455  3588
                   4820   5042      5153  5350      5529  5889      6318  6426
                   6576   6634      6723  6792
  1.969 11.174     1579   1663      2540  2635      3455  3588      4820  5042
                   5153   5350      5529  5889      6318  6426      6576  6634
                   6723   6792

...
```

GeneModeler (GM)

Reference   Soderlund *et al.*, 1992
FTP   `ftp.tigr.org` or `atlas.lanl.gov`, directory/pub/gm

Description

GM is a gene identification program for Unix platforms. It has been specifically tuned to suggest possible gene structures in *C. Elegans* sequences, but can also be used to analyze sequences from other organisms. GM consists of a set of pattern matching routines, and a main engine that assembles exon maps based on the analysis of such routines. In the first step, ORFs and potential coding regions are identified. Then, candidate splice sites are identified within such potential coding regions. In frame groups of candidate exons are then placed in gene maps. The implied introns, and gene beginning and end are next evaluated. Maps passing all of the tests are searched for an optional upstream promoter site, which is added to the model if found. GM also allows for exon maps to be extended from known cDNAs. Two important strengths of GM are a flexible graphical user interface (allowing the display of multiple possible gene structures and their potential amino acid translations, along with the nucleotide sequence) and easy replacement of program parameters by the user (allowing customization to new organisms).

GeneParser

Reference   Snyder and Stormo, 1995a
Inquiries   `eesnyder@sequana.com`
FTP   `beagle.colorado.edu`, directory /pub/GeneParser

WWW   http://beagle.colorado.edu/~eesnyder/
           GeneParser.html

Description

GeneParser identifies coding exons in genomic human DNA. The program scores all subintervals in the query sequence for content statistics (in-frame and bulk hexamers, local compositional complexity, length,...) indicative of introns and exons, and for sites that identify boundaries. The information is weighted by a neural network to approximate the log-likelihood that each subinterval exactly represents an intron or exon. A dynamic programming algorithm is then applied to find the combination of introns and exons that maximizes the likelihood function. Among the content statistics computed on the sequence subintervals, GeneParser allows the option of computing the blast similarity scores of the subinterval against a protein sequence database. GeneParser is unusual in that it does not filter out any poor scoring exon candidates, but considers every possible combination of potential exons in searching for a potential solution. This thoroughness comes at the cost, of course, of longer execution time. In addition, GeneParser does not enforce open reading frame constraints, thus it may be very appropriate to analyze data likely to contain sequencing errors. GeneParser also has a sophisticated graphical user interface that aids the user to explore alternate gene models.

GenLang

Reference   Dong and Searls, 1994
Email   send message "help" to genlang@cbil.humgen.upenn.edu
FTP   cbil.humgen.upenn.edu directory/pub/genlang
WWW   http://cbil.humgen.upenn.edu/~sdong/
            genlang_home.html

Description

GenLang takes a linguistic approach, considering genes to be the "sentences" of a "language" defined by a formal grammar, and presents a grammatical parse of the sequence. GenLang's grammar specifies how any gene or component of a gene must be made up of smaller constituents. For example one grammar rule states that an intron must begin with a donor site and end with an acceptor. The most detailed rules state how to evaluate direct evidence from the sequence; for example, how to score a putative acceptor site using a weight matrix, a coding measure value upstream, and one downstream. At each point where constituents are combined, a rule is given for combining the constituent scores. When GenLang is "trained" on sequences from a particular organism, these rules are refined to reflect the importance of each kind of evidence in this particular context. There are general-purpose parsers that can find all possible interpretations of a sequence in accord with a particular grammar. GenLang applies one of these but, for efficiency, aborts any parse with a very low-scoring constituent. The parser is allowed to run for a specified time, and then the best-scoring parse found is reported. Because of the explicit modeling of gene structure, it seems likely that GenLang will be one of the first programs to incorporate an improved understanding of transcription and splicing.

Using GenLang

One sequence can be sent to the server in a single Email message (genlang@cbil.humgen.upenn.edu). Four lines should precede the sequence in the body of the message. The first one contains the NAME of the sequence, the second line, the SPECIES. Three taxonomic groups are currently supported: drosophila, vertebrates and dicots. The third line, the type of GENE. Besides protein coding genes, GenLang can also search for tRNA genes, and group I introns. The fourth line should contain the keyword SEQUENCE, and the following lines the sequence itself. Below is a sample message.

```
NAME Human enolase 3
SPECIES vertebrate
GENE protein
SEQUENCE
AGATCTCTACCGAGGGCAGAGACCTACCTCCCCGCAGTGCTACAAGTGGGGCGCCGGAAG
AGCCCCAGGCGTGCAGAAGCTCACAAAAGGCCACCCGTCCTCGGTCCATTCATTTTTGTT
CACTGTTGATTCAGCCCCATTCATTGATGGGCTGGGGCCGTGCGCTGAGCGCCCACAGTC
...
```

The output consists of up to the 10 lowest cost gene structures. The output for this sequence (trimmed) is:

```
Length of the sequence: 7,194

The 10 best (lowest cost) gene structures found by GenLang are:

Rank=3, Cost=88
(1579/1663,4364/4476,4820/5042,5153/5350,5529/5889,6318/6426,6576/6683)
...
 Rank=6, Cost=91
(1579/1663,4364/4476,4820/5042,5153/5350,5529/5889,6723/6792)
...
```

The WWW server may be used in place of Email for communication with the server. The output produced is the same as that produced by the Email server, and is returned as an html document. In addition, a postsript illustration can be obtained of the best parses found by GenLang.

GRAIL

Reference Xu *et al.*, 1994
Email   send message "help" to grail@ornl.gov
FTP   arthur.epm.ornl.gov, in directory pub/xgrail
WWW   http://avalon.epm.ornl.gov/gallery.html

Description

GRAIL is a suite of tools aimed at the recognition of coding regions in DNA sequences and the prediction of gene structure in human sequences. The GRAIL system provides three types of analysis, recognition of coding regions (GRAIL 1, 1a), prediction of exons (GRAIL 2), and assembly of genes (GAP 3). GRAIL 1 uses a neural network to integrate information from seven coding measures on a fixed size sequence window. On a query sequence, the input values are computed on windows centered at successive positions 10 bp apart. If the window's coding potential is above a pre-fixed threshold, the position is considered to be coding. GRAIL 1a, in addition, re-evaluates the endpoints of potential coding regions, using information from the two 60-base regions adjacent to each, to find the "best" boundaries.

GRAIL 2 predicts complete coding exons, with splicing boundaries. GRAIL 2 divides the exon recognition process into four main steps. In the first step, all possible initial, internal, and terminal exons are generated. In the second step, candidate exons are filtered out using a set of heuristic rules that each probable candidate exon must satisfy. In the third step, remaining exons are evaluated and scored by a neural networks specific to the type of exon (initial, internal, and terminal), using coding measures (statistical tests for regularities typical of coding regions) and values related to the quality of the end points. In the final step, exons are clustered on the basis of location, and the best scoring candidate taken from each cluster. GAP III assembles gene models for a query sequence from the set of clusters of candidate exons generated by GRAIL 2. GAP III uses a dynamic programming algorithm to pick up at most one exon for each cluster—the exon picked up may not be the higher scoring one within the cluster—to produce the highest scoring gene model. The algorithm is designed so that exons are included in the predicted gene model from as many clusters as possible.

GRAIL 1 may be used to analyze not only genomic sequences, but also short DNA fragments from single shotgun sequencing gels, and cDNAs. GRAIL 2 and GAP III, however, are not appropriate for sequences without genomic context. The GRAIL suite of tools is constantly under development, with new techniques being rapidly incorporated, and users may expect a high level of service. For independent evaluations of GRAIL, see Claverie, 1994b; Lopez *et al.*, 1994; Singh and Krawetz, 1994. The GRAIL tools can also be accessed through the XGRAIL system. XGRAIL is an X-based Internet client server implementation of GRAIL. The client runs the front-end graphical interface locally, while most of the sequence analysis is performed remotely at the server.

Using Grail

We illustrate GRAIL 2 here, and GRAIL 1 in the next section; GAP III is not available by Email. Any number of sequences can be sent to the server (grail@ornl.gov) in a single Email message. The first line of the sample message below (beginning with "Sequences") specifies the kind of analysis desired. The first number means that there is only one sequence in the message; the "-2" requests GRAIL 2; and the "-E" requests database comparisons of the translations of exons with an "excellent" rating.

```
Sequences 1 -2 -E
> Human enolase 3
AGATCTCTACCGAGGGCAGAGACCTACCTCCCCGCAGTGCTACAAGTGGGGCGCCGGAAG
AGCCCCAGGCGTGCAGAAGCTCACAAAAGGCCACCCGTCCTCGGTCCATTCATTTTTGTT
CACTGTTGATTCAGCCCCATTCATTGATGGGCTGGGGCCGTGCGCTGAGCGCCCACAGTC
...
```

The GRAIL 2 output in this case:

```
> Human enolase 3, len = 7194
```

Exon predication on forward strand:

start/acceptor		donor/stop	rf	score	orf		
1579	–	1663	1	81	1459	-	1686
2540	–	2635	1	87	2437	-	2973
2796	–	2858	2	41	2762	-	3022
3455	–	3588	1	91	3400	-	3888
4820	–	5042	2	92	4706	-	5053
5153	–	5350	1	92	5041	-	5412
5688	–	5889	2	80	5522	-	5893
6576	–	6648	3	46	6549	-	6683
6723	–	6788	1	88	6706	-	6792

Exon predication on reverse strand:

start/donor		acceptor/stop	rf	score	orf		
5548	–	5632	2	47	5330	-	5638
2095	–	2417	3	38	2070	-	2417

Final Exon Predication:

start/donor		acceptor/stop	strand	rf	quality	orf		
1579	–	1663	f	1	excellent	1459	-	1686
2540	–	2635	f	1	excellent	2437	-	2973
2796	–	2858	f	2	good	2762	-	3022
3455	–	3588	f	1	excellent	3400	-	3888
4820	–	5042	f	2	excellent	4706	-	5053
5153	–	5350	f	1	excellent	5041	-	5412
5688	–	5889	f	2	excellent	5522	-	5893
6576	–	6648	f	3	good	6549	-	6683
6723	–	6788	f	1	excellent	6706	-	6792

SORFIND

Reference   Hutchinson and Hayden, 1993

Inquiries   Gordon B. Hutchinson at `hutch@netshop.bc.ca`

Description

SORFIND is a program for IBM PC compatibles that predicts internal exons in human genomic sequences. A more recent version (the extensions are not described here) also predicts initial and terminal exons. The program searches a sequence for open reading frames bracketed by potential splice sites, called "spliceable open reading frames" or SORFs. SORFs are scored using a linear function that combines splice site scores (based on an energy calculation as discussed above) and three coding measures. At this step, overlapping SORFs or SORFs being less than a minimum distance (minimum intron length) away are clustered into a single group, and the best candidate within each group is chosen. Following this, the sequence is partitioned, so that further analysis will avoid regions already containing a successful SORF and its surrounding minimum introns. An important strength of the program is that potential exons are predicted with carefully defined and quantitatively interpretable confidence levels.

Coding Region Analysis of Sequence Fragments

Coding region analysis involves subjecting successive "windows" of the sequence, usually about 100 bases long each, to statistical tests for regularities typical of coding regions. One can expect roughly 85-90% accuracy on fully coding or noncoding windows (Fickett and Tung, 1992), with somewhat unpredictable behaviour at the endpoints of coding regions. Two recent papers (Claverie, 1994b; Kamb *et al.*, 1995) analyze the likelihood of missing a gene when first-pass shotgun sequences are subjected to a coding region analysis.

GeneMark

Reference   Borodovsky and McIninch, 1993
Email   send message "help" to genemark@ford.gatech.edu
             or genemark@embl-ebi.ac.uk

Description

GeneMark is a coding region and exon detection program. It is best known for its ability to analyze *E. coli* sequences, but has been extended to analyze sequences from a wide variety of organisms, both prokaryotic and eukaryotic. GeneMark has seven probability (Markov chain) models for sequences that are presented in each of six coding frames, or are noncoding. The probability models for coding DNA are based on counts of in-phase hexamer frequencies (though the word length—6 for hexamers—may vary depending on the organism). The algorithm evaluates the probability that the given sequence was produced under each of these seven models, then uses Bayes' theorem to calculate the probability of each model being the correct one. This analysis of coding regions may also be combined with a search for start and stop codons, or a simple search for intron junctions. A feature of GeneMark is that, distinct to other methods, identification of coding regions is performed simultaneously in both strands.

The service has seen extensive practical use, for example in the *E. coli* genome sequencing project, and is applicable to both genomic and cDNA sequences of a number of organisms.

Using GeneMark

A human sequence is sent to GeneMark as follows:

```
To: genemark@ford.gatech.edu
Subject: genemark

species human
data
AGATCTCTACCGAGGGCAGAGACCTACCTCCCCGCAGTGCTACAAGTGGGGCGCCGGAAG
AGCCCCAGGCGTGCAGAAGCTCACAAAAGGCCACCCGTCCTCGGTCCATTCATTTTTGTT
CACTGTTGATTCAGCCCCATTCATTGATGGGCTGGGGCCGTGCGCTGAGCGCCCACAGTC
...
```

GeneMark's basic output is an ORF chart. For eukaryotic sequences, a list of potential protein coding exons is also provided.

```
                         GENEMARK PREDICTIONS

Query:
Sequence file: /tmp/caaa0079S
Sequence length: 7194
GC Content:   55.09%
Window length: 96
Window step: 12
Threshold value: 0.500
-
Matrix: H. sapiens, 0.52 < GC < 0.58 - Order 4
Matrix author: JDM
Matrix order: 4

List of Regions of interest
(regions from stop tp stop codon w/ coding function >0.500000)
```

LEnd	REnd	Strand	Frame
1456	1686	direct	fr 1
2434	2973	direct	fr 1
3397	3888	direct	fr 1
4703	5053	direct	fr 2
5038	5412	direct	fr 1
5519	5893	direct	fr 2

```
List of Open reading frames
(regions from start to stop codon w/ coding function >0.500000)
```

LEnd	REnd	Strand	Frame	Prob	Start
1579	1686	direct	fr 1	0.5448	0.9946
1585	1686	direct	fr 1	0.5778	0.9943
4829	5053	direct	fr 2	0.9051	0.9994
4868	5053	direct	fr 2	0.8902	0.9998

```
4880      5053   direct          fr 2   0.8823   0.9999

5047      5412   direct          fr 1   0.5208   0.0000
5203      5412   direct          fr 1   0.7661   1.0000
5206      5412   direct          fr 1   0.7661   1.0000
5215      5412   direct          fr 1   0.7515   0.9999

5687      5893   direct          fr 2   0.7218   0.9915
5753      5893   direct          fr 2   0.7270   1.0000
5768      5893   direct          fr 2   0.6998   1.0000
```

List of Protein-Coding Exons
(regions between acceptor and donor site w/ coding function >0.500000)

Left End	Right End	Strand	Frame	Prob
1223	1282	direct	fr 1	0.7019
1575	1636	direct	fr 1	0.7818
2563	2636	direct	fr 1	0.8425
2805	2859	direct	fr 2	0.6222
3453	3583	direct	fr 1	0.5336
3458	3534			0.6864
3935	4000	direct	fr 1	0.3791
4798	5007	direct	fr 2	0.8605
4818	4998			0.9625
5159	5351	direct	fr 1	0.9053
5168	5322			0.9645
5701	5838	direct	fr 2	0.9438
5715	5816			0.9846
6332	6403	direct	fr 1	0.4239

In addition, GeneMark can send graphical output of its results in the form of a PostScript file. It is essentially the plot of the coding probability function along the sequence for each of the six frames, and also indicates potential start and stop codons and splice sites.

GRAIL

GRAIL is a suite of tools aimed at the recognition of coding regions in DNA sequences and the prediction of gene structure in human sequences. See above.

Using GRAIL 1

To use GRAIL 1 rather than GRAIL 2, one simply leaves out the "-2" option on the "Sequences" line, so:

```
To: grail@ornl.gov
Subject:

Sequences 1 -E
> Human enolase 3
AGATCTCTACCGAGGGCAGAGACCTACCTCCCCGCAGTGCTACAAGTGGGGCGCCGGAAG
AGCCCCAGGCGTGCAGAAGCTCACAAAAGGCCACCCGTCCTCGGTCCATTCATTTTTGTT
CACTGTTGATTCAGCCCCATTCATTGATGGGCTGGGGCCGTGCGCTGAGCGCCCACAGTC
```

The program tests successive 100 base windows, stepping 10 bases at a time, for coding potential, and calculates a score between 0 and 1 for each strand of each window. A score over 0.5 indicates likely coding function. To reduce the volume of output, only windows for which at least one strand had a score of at least 0.01 are listed. For the above sequence:

```
> Human enolase 3, len = 7194
```

pos	f-strand	frame	orf		pos	r-strand	frame	orf
131	0.000	-	- - -	||	7061	0.085	-	- - -
141	0.000	-	- - -	||	7051	0.165	-	- - -
351	0.074	-	- - -	||	6841	0.000	-	- - -
941	0.011	-	- - -	||	6251	0.145	-	- - -
951	0.123	-	- - -	||	6241	0.117	-	- - -
961	0.246	-	- - -	||	6231	0.147	-	- - -
1181	0.084	-	- - -	||	6011	0.000	-	- - -
1191	0.409	-	- - -	||	6001	0.000	-	- - -
1201	0.937	1	1165 - 1330	||	5991	0.000	-	- - -
1211	0.619	1	1165 - 1330	||	5981	0.000	-	- - -
1221	0.877	1	1165 - 1330	||	5971	0.000	-	- - -
1231	0.191	-	- - -	||	5961	0.000	-	- - -
1241	0.341	-	- - -	||	5951	0.000	-	- - -
1251	0.116	-	- - -	||	5941	0.000	-	- - -
1261	0.077	-	- - -	||	5931	0.000	-	- - -
1591	0.230	-	- - -	||	5601	0.007	-	- - -
1601	0.938	1	1456 - 1684	||	5591	0.021	-	- - -
1611	0.938	1	1456 - 1684	||	5581	0.081	-	- - -

...

Xpound

Reference   Thomas and Skolnick, 1994
Inquiries   Alun Thomas at alun@myriad.com

Description

Xpound is a program for Unix platforms that predicts coding regions, with frame, in DNA sequences. Xpound is based on a probabilistic model. The nucleotides in a sequence are considered as belonging to seven different classes: those in the three codon positions, those in intergenic regions, and those that are in introns breaking the coding sequence at each of the three possible codon positions. Two simplifying assumptions are made: (1) that the probability of encountering one of these classes of nucleotides depends only on the class of the upstream nucleotide, and (2) that the probability distribution of the bases at each position of the sequence depends only on the bases and classes of bases in the immediate vicinity. Xpound then uses Bayes's theorem to make a maximum likelihood estimate of the state at each base of the query sequence. A strength of Xpound is that it is very easy to reset all the relevant probabilities for use with sequences of a new organism.

Signal Identification

Some signal identification techniques have been developed further in isolation than within the context of integrated gene identification algorithms. Thus it may be of value to make use of the following services as well.

Netgene

Reference   Brunak *et al.*, 1991
Email   send message "help" to netgene@cbs.dtu.dk

Description

NetGene is a neural network method that predicts splice site locations in human pre-mRNA. Neural networks are used to score nucleotides in the query sequence as donor, and acceptor sites. The input to the neural networks is a sequence window centered in the scored nucleotide. The cutoff level for splice site assignment is not constant over the query sequence, but is regulated by the prediction of transition regions between introns and exons. Thus, in regions with abrupt change in predicted coding behavior, splice sites can be predicted even with relatively low score, while in regions with only small change in predicted coding activity, a rather high confidence level in the splice site assignment is demanded. Transitions between introns and exons are predicted by computing the derivative of the output of a separate neural network trained to recognize coding exons.

A sample request to the NetGene server:

```
To: netgene@virus.fki.dth.dk
Subject:

> Human enolase 3
AGATCTCTACCGAGGGCAGAGACCTACCTCCCCGCAGTGCTACAAGTGGGGCGCCGGAAG
```

AGCCCCAGGCGTGCAGAAGCTCACAAAAGGCCACCCGTCCTCGGTCCATTCATTTTTGTT
CACTGTTGATTCAGCCCCATTCATTGATGGGCTGGGGCCGTGCGCTGAGCGCCCACAGTC
. . .

And the (trimmed) results (note the more confident predictions in a separate section first):

```
----------------------------------------------------------------
The sequence: Human enolase 3 contains 7194 bases, and has the following
composition:
A 1553 C 1992 G 1971 T 1678
```

1) HIGHLY CONFIDENT SITES:
`===========================`

ACCEPTOR SITES:

POSITION	CONFIDENCE	INTRON EXON	LOCAL	GLOBAL
3454	0.42	CTCCTTCCAG^CCAAGTTTGG	0.84	0.62
4819	0.42	CCCATCTCAG^GCCTTCAATG	0.72	0.72
1576	0.37	TGTCCTGCAG^CCATGGCCAT	0.70	0.69
5152	0.20	CCTCCCCCAG^CCCTGGAGCT	0.61	0.62
3015	0.04	CACCTTCCAG^AAACTAAGCG	0.56	0.53

DONOR SITES:

POSITION	CONFIDENCE	EXON INTRON	LOCAL	GLOBAL

2) NEARLY ALL TRUE SITES:
`===========================`

ACCEPTOR SITES:

POSITION	CONFIDENCE	INTRON EXON	LOCAL	GLOBAL
3454	0.73	CTCCTTCCAG^CCAAGTTTGG	0.84	0.62
4819	0.67	CCCATCTCAG^GCCTTCAATG	0.72	0.72
1576	0.64	TGTCCTGCAG^CCATGGCCAT	0.70	0.69
5152	0.51	CCTCCCCCAG^CCCTGGAGCT	0.61	0.62
6121	0.46	CCTCTCTCAG^ATTCCGCTGT	0.80	0.27
1771	0.42	CCTTTCCCAG^ACTTCTTCCC	0.67	0.40
3015	0.40	CACCTTCCAG^AAACTAAGCG	0.56	0.53
515	0.34	AGCCTTGCAG^GGCTCAGGTT	0.77	0.13
2539	0.33	CCTGTCCCAG^GCCGATTCCG	0.71	0.19

. . .

DONOR SITES:

POSITION	CONFIDENCE	EXON INTRON	LOCAL	GLOBAL
704	0.27	TTGACCTTTG^GTAAGGGGGC	0.98	0.10
6318	0.27	CTCCACTCAG^GTGCAAACTG	0.91	0.19
379	0.26	ACCCTAAGAG^GTGAGACCCT	0.91	0.18
3589	0.25	CCCAGTGCCA^GTGAGTGCAG	0.60	0.53
1664	0.22	ACGGCCAAGG^GTAACACAAG	0.41	0.73

. . .

```
----------------------------------------------------------------
```

PromoterScan

Reference   Prestridge, 1995.
Inquiries   Dan Prestridge at `danp@biosci.cbs.umn.edu`

Description

PromoterScan is a program (available for Unix and IBM-compatibles) that predicts PolII transcription start sites in eukaryotic sequences. In the calibration step, consensus sequences are used to recognize putative transcription factor binding sites in a training set of promoter and non-promoter regions, and ratios of densities for putative binding sites in promoters and non-promoters are recorded for all transcription factors in the Transcription Factor Database (Ghosh, 1990). In application, the density ratios of putative transcription factor binding sites (again recognized by means of consensus sequences) are summed, and this score is combined with the Bucher weight matrix score (Bucher, 1990) of any putative TATA box. When the score threshold is set so that 70% of promoters are recognized correctly, one false positive is recorded about once every 5600 bases.

Using PromoterScan

Sending an Email message to Dan Prestridge at danp@biosci.cbs.umn.edu resulted in clear instructions for retrieval and installation of the program. The sequence was supplied to PromoterScan naked:

```
AGATCTCTACCGAGGGCAGAGACCTACCTCCCCGCAGTGCTACAAGTGGGGCGCCGGAAG
AGCCCCAGGCGTGCAGAAGCTCACAAAAGGCCACCCGTCCTCGGTCCATTCATTTTTGTT
CACTGTTGATTCAGCCCCATTCATTGATGGGCTGGGGCCGTGCGCTGAGCGCCCACAGTC
...
```

The resulting predictions were:

```
Predicted promoter in hseno3.seq from (+)205-455
Predicted promoter in hseno3.seq from (+)609-859, Predicted TATA BOX at
(+)835
Predicted promoter in hseno3.seq from (+)949-1199, Predicted TATA BOX at
(+)1149
Predicted promoter in hseno3.seq from (-)3367-3617, Predicted TATA BOX at
(-)3390
Predicted promoter in hseno3.seq from (+)3632-3882
Predicted promoter in hseno3.seq from (+)4520-4770
```

Acknowledgments

This work was supported by Public Health Service grant #HG00981-01A1 from the National Center for Human Genome Research, and by the Fundació Catalana per a la Recerca.

References

Adams, M.D., Fields, C. and Venter, J.C. 1994. Automated DNA sequencing and analysis. Academic Press, San Diego, CA.

Altman, R., Brutlag, D., Karp, P., Lathrop, R. and Searls, D.1994. Proceedings of The Second International Conference on Intelligent Systems for Molecular Biology.

AAAI Press, Menlo Park, CA.

Altschul, S.F., Gish, W., Miller, W., Myers, E.W. and Lipman, D.J. 1990. Basic Local Alignment Search Tool. J. Mol. Biol. 215: 403-410.

Bairoch, A. 1992. PROSITE: A dictionary of sites and patterns in proteins, Nucl. Acids Res. 20: 2013-2018.

Borodovsky, M. and McIninch, J. 1993. GENMARK: Parallel gene recognition for both DNA strands. Computers in Chemistry. 17: 123-134.

Brunak, S., Engelbrecht, J. and Knudsen, S. 1991. Prediction of human mRNA donor and acceptor sites from the DNA sequence. J. Mol. Biol. 220: 49-65.

Bucher, P. 1990. Weight matrix descriptions of four eukaryotic RNA polymerase II promoter elements derived from 502 unrelated promoter sequences. J. Mol. Biol. 212: 563-578.

Califano, A. and Rigoutsos, I. 1993. FLASH: A Fast Look-up Algorithm for String Homology. In: Proceedings of the First International Conference on Intelligent Systems for Molecular Biology, July 1993, Bethesda, MD.

Claverie, J.-M. 1992. Identifying coding exons by similarity search: Alu-derived and other potentially misleading protein sequences. Genomics. 12: 838-841.

Claverie, J.-M. 1994a. Large scale sequence analysis. In: Automated DNA Sequencing and Analysis. M.D. Adams, C. Fields, and J.C. Venter, eds. Academic Press, San Diego, CA. pp267-279

Claverie, J.-M. 1994b. A streamlined random sequencing strategy for finding coding exons. Genomics. 23: 575-581.

Dong, S. and Searls, D.B. 1994. Gene structure prediction by linguistic methods. Genomics. 23: 540-551.

Fickett, J.W. and Tung, C.-S. 1992. Assessment of protein coding measures. Nucl. Acids Res. 20: 6441-6450.

Gelfand, M.S. 1995. Prediction of function in DNA sequence analysis. J. Comp. Biol. 2: 87-115.

Ghosh, D. 1990. A relational database of transcription factors. Nucl. Acids. Res. 18: 1749-1756.

Green, P., Lipman, D., Hillier, L., Waterston, R., States, D. and Claverie, J.-M. 1993. Ancient conserved regions in new gene sequences and the protein databases. Science. 259: 1711-16.

Gribskov, M. and Devereux, J. 1991. Sequence Analysis Primer. Stockton Press. New York, NY.

Guigó, R., Knudsen, S., Drake, N. and Smith, T. 1992. Prediction of gene structure. J. Mol. Biol. 226: 141-157.

Henikoff, S. and Henikoff, J.G. 1994. Protein family classification based on searching a database of blocks. Genomics. 19: 97-107.

Hutchinson, G.B. and Hayden, M.R. 1993. SORFIND: A computer program that predicts exons in vertebrate genomic DNA. In: Proceedings Of The Second International Conference On Bioinformatics, Supercomputing, and Complex Genome Analysis. H.A Lim, J.W. Fickett, C.R. Cantor and R.J. Robbins, eds. World Scientific, Singapore. pp 513-520.

Kamb, A., Wang, C., Thomas, A., DeHoff, B.S., Norris, F.H., Richardson, K., Rine, J., Skolnick, M. and Rosteck, P.R. Jr. 1995. Software trapping: A strategy for finding genes in large genomic regions. Computers and Biomed. Res. 28: 140-153.

Koonin, E.V., Bork, P. and Sander, C. 1994. Yeast chromosome III: New gene functions. EMBO J. 13: 493-503.

Krogh, A., Mian, I.S. and Haussler, D. 1994. A hidden Markov model that finds genes in *E. coli* DNA. Nucl. Acids Res. 22: 4768-4778.

Krogh, A., Mian, I.S. and Haussler, D. 1994. A hidden Markov model that finds genes in *E. coli* DNA. UCSC Report CRL-93-33.

Lipmann, R.P. 1987. An Introduction to Computing with Neural Nets. IEEE ASSP Magazine. April: 4-22.

Lopez, R., Larsen, F. and Prydz, H. 1994. Evaluation of the exon predictions of the GRAIL software. Genomics. 24: 133-136.

Milanesi, L., Kolchanov, N.A., Rogozin, I.B., Ischenko, I.V., Kel, A.E., Orlov, Y.L., Ponomarenko, M.P. and Vezzoni, P. 1993. GenViewer: A computing tool for protein-coding regions prediction in nucleotide sequences. In: Proceedings of the Second International Conference on Bioinformatics, Supercomputing, and Complex Genome Analysis. H.A. Lim, J.W. Fickett, C.R. Cantor and R.J. Robbins, eds. World Scientific, Singapore. pp 573-587.

Milosavljevic, A. 1995. Repeat analysis. In: Imperial Cancer Research Fund Handbook of Genome Analysis. Blackwell Scientific Publications, Oxford, UK. Chapter 13, section 4 (In press).

Ogiwara, A., Uchiyama, I., Seto, Y. and Kanehisa, M. 1992. Construction of a dictionary of sequence motifs that characterize groups of related proteins. Protein Eng. 5: 479-488.

Pearson, W.R. and Lipman, D.J. 1988. Improved tools for biological sequence comparison. Proc. Natl. Acad. Sci. U.S.A. 85: 2444-2448.

Prestridge, D.S. 1995. Predicting Pol II promoter sequence using transcription factor binding sites. J. Mol. Biol. 249: 923-932.

Singh, G.B. and Krawetz, S.A. 1994. Computer based exon detection: An evaluation metric for comparison. Internat. J. Genome Res. 1: 321-338.

Snyder, E.E. and Stormo, G.D. 1995a. Identification of protein coding regions in genomic DNA. J. Mol. Biol. 248: 1-18.

Snyder, E.E. and Stormo, G.D. 1995b. Identifying genes in genomic DNA sequences. In: DNA and Protein Sequence Analysis: A Practical Approach. M.J. Bishop and C.J. Rawlings, eds. IRL Press, Oxford, England (In press).

Soderlund, C., Shanmugam, P., White, O. and Fields, C. 1992. In: Proceedings of The 25th Hawaii International Conference on System Sciences, V. Milutinovic, V. and Shriver B.(Eds.). IEEE Computer Society Press, Los Alamitos, CA. pp 653-662.

Solovyev, V.V., Salamov, A.A. and Lawrence, C.B. 1994. The prediction of human exons by oligonucleotide composition and discriminant analysis of spliceable open reading frames. In: Proceedings of the Second International Conference on Intelligent Systems for Molecular Biology. R. Altman, D. Brutlag, P. Karp, R. Lathrop and D. Searls, eds. AAAI Press, Menlo Park, CA. pp 354-362.

Sturrock, S.S. and Collins, J.F. 1993. MPsrch version 1.3. Biocomputing Research Unit, University of Edinburgh, UK.

Thomas, A. and Skolnick, M.H. 1994. A probabilistic model for detecting coding regions in DNA sequences. IMA J. of Math. Applied in Med. and Biol. 11: 149-160.

Williamson, A.R., Elliston, K.O. and Sturchio, J.L. 1995. The Merck Gene Index: A Public Resource for Genomics Research. Journal of NIH Research. 7: 61-64

Xu, Y., Einstein, J.R., Mural, R.J., Shah, M. and Uberbacher, E.C. 1994. An improved system for exon recognition and gene modeling in human DNA sequences. In: Proceedings of The Second International Conference on Intelligent Systems for Molecular Biology. R. Altman, D. Brutlag, P. Karp, R. Lathrop and D. Searls, eds. AAAI Press, Menlo Park, CA. pp 376-383.

Further Reading

Barrick, D., Villanueba, K., Childs, J., Kalil, R., Schneider, T.D., Lawrence, C.E., Gold, L. and Stormo, G.D. 1994. Quantitative analysis of ribosome binding sites in *E. coli*. Nucl. Acids Res. 22: 1287-1295.

Doolittle, R.F. 1990. Molecular Evolution: Computer Analysis of Protein and Nucleic Acid Sequences. Vol 183.

Gelfand, M.S. and Roytberg, M.A. 1993. A dynamic programming approach for prediction of the exon-intron structure. Bio Systems. 30: 173-182.

Lim, H.A., Fickett, J.W., Cantor, C.R. and Robbins, R.J. 1993. Proceedings of the Second International Conference on Bioinformatics, Supercomputing, and Complex Genome Analysis. World Scientific. Singapore.

Stormo, G.D. 1990. Consensus patterns in DNA. In: Molecular Evolution: Computer Analysis of Protein and Nucleic Acid Sequences, Methods in Enzymology. 183: 211-220.

von Hippel, P.H. 1994. Protein-DNA recognition: new perspectives and underlying themes. Science. 263: 769-770.

From: *Internet for the Molecular Biologist.*
ISBN 1-898486-02-6 ©1996 Horizon Scientific Press, Wymondham, U.K.

9

THE BIOSCI/BIONET ELECTRONIC NEWSGROUP NETWORK FOR BIOLOGISTS

David Kristofferson

Introduction

Beginning in late 1993, the Internet started making headlines in the mass media, primarily because of the advent of the Mosaic World Wide Web (WWW) browser produced at the University of Illinois' National Center for Supercomputing Applications. Despite the headlines caused by WWW, when this author polled biologists on several recent seminar tours, it was plain to see that electronic mail was still the dominant Internet application used by research biologists. The reason for this is simple. The Internet is a communications tool. E-mail is the most widely available communications application, included as a standard software feature on most computer systems beyond the desktop level. Gopher, USENET news software, Mosaic, etc., usually have to be installed after a computer is up and running, and this sometimes presents a barrier for scientists if they don't have good local systems support. However, virtually everyone with an account on a networked computer can access Email, the only barrier being the self-imposed one of resistance to learning how to use it. Knowledge of how to send, read, and reply to Email is therefore our assumed foundation for the discussion that follows.

Chris Goes Surfing

To relieve a bit of the tedium that biologists might feel when reading about computing, let's personalize this chapter on electronic newsgroups for biologists by relating the progress of a mythical scientist named Chris, a good unisex name, as he/she discovers Email and the communication possibilities for biologists on the net. Having once mastered Email on the local computer system, Chris begins sending messages to colleagues and then learns how to insert files in mail messages. This allows Chris to transmit data and even manuscripts. For some time after learning these techniques, this is the extent of Chris' network explorations - there's simply too much lab work to be done and journal articles to read.
     One day while browsing "Trends in Biochemical Sciences", Chris chances upon Paul Hengen's monthly "Methods and Reagents" column in the "Computer Corner" section. The article fortunately happens to contain an important insight into

an experimental problem that Chris has been struggling with. Chris reads further and finds that the source of the information is an Internet newsgroup system called BIOSCI. The article says that BIOSCI instructions can be retrieved automatically by sending an electronic mail message to the Internet address.

```
biosci@net.bio.net.
```

The message does not require any special contents or syntax, so, since this sounds pretty simple to do, Chris fires off a blank Email message to the address. Not too long after sending the message, Chris receives the automatic reply from the BIOSCI Email server. The message notes that BIOSCI is a set of electronic "USENET newsgroups" and mailing lists for biologists (readers please note that the list of BIOSCI newsgroups is growing each month and is not included here since it would be rapidly out of date - please get the latest one from the net). Chris scans the list and sees that, in addition to the METHODS-AND-REAGENTS mailing list, there are about 80 topics available, several of which are of potential interest. Not only are there general lists for scientific employment ads, research journal table of contents postings (ahead of publication!), and information about scientific software, Chris also reads about a large number of lists dedicated to various areas of biology research which allow people from all over the world to discuss problems/issues with each other. Best of all, access to all of these services is *free*.

The only catch is that the message rambles on about how participants in BIOSCI should use USENET news instead of Email. Since Chris hasn't really heard of USENET news before, much less has any desire to spend additional time learning more computer programs, Chris skims the message until finding the section on how to sign up by Email. It turns out that there are two BIOSCI Email sites, one in the U.S. at a computer called net.bio.net which handles Email distribution for users in the Americas and Pacific Rim countries and another in the U.K. which handles Email distribution for users in Europe, Africa, and central Asia. Each site has an address for personal technical support.

Support Address	**For users in**
`biosci@daresbury.ac.uk`	Europe, Africa, and Central Asia
`biosci-help@net.bio.net`	Americas and the Pacific Rim

The message also describes how to retrieve instructions for adding one's Email address to the BIOSCI mailing lists automatically. Being located in the U.S., Chris sends in the command:

```
lists
```

to the address `biosci-server@net.bio.net` to retrieve the set of mailing list names that the net.bio.net Email server recognizes. Chris peruses this list and then sends in the final subscription message. The message is addressed to `biosci-server@net.bio.net` and the Subject: line is left blank (anything entered on it is ignored). Chris decides to subscribe to several lists by putting the following commands in the mail message:

```
subscribe bionews
subscribe bioforum
subscribe celegans
subscribe protista
subscribe xtal-log
subscribe womenbio
subscribe jrnlnote
subscribe microbio
end
```

A few minutes later the Email server sends back acknowledgements that Chris' subscriptions have been entered and also sends some additional information on the lists requested. As it is getting a bit late on Friday evening, Chris decides to call it quits. Due to the press of lab work, Chris doesn't get a chance to login in to the computer again until the following Wednesday. Unfortunately, Chris is annoyed to find that there are a couple of hundred messages in the Email file and belatedly understands why the BIOSCI instructions did not recommend the use of Email. Nonetheless, sheer determination wins out and Chris finds that there actually were several messages of interest among the 231 that had to be deleted. Reflecting on the traditional literature, Chris realizes that no one reads journals from cover to cover either, but instead selects articles from a table of contents. If only there was a table of contents for all of this stuff!

Rereading the BIOSCI instructions, Chris realizes that this is what USENET news software does. It sorts messages by newsgroup and, within each newsgroup, by discussion topic, thereby creating a "Table of Contents" out of the message Subject: lines. Chris checks with the campus computer center and discovers that a news program is in fact available on the campus computer. It takes a little time to learn how to use, and the local systems administrator, who is always a bit harried, doesn't help Chris out much in learning to use it. Nevertheless Chris sees that learning the news program is not much different from learning to use an Email program, and promises to save a lot of time compared to continued participation in BIOSCI by Email. News software simply sorts messages more effectively, saves them in a central location for all users of the computer to access (instead of cluttering up everyone's mail files), and allows the user to participate in BIOSCI by reading and posting to the "bionet" newsgroups in the newsreader. The newsreader contains the same messages that were also sent to Chris' Email address, so Chris sends one last message to `biosci-server@net.bio.net`:

```
unsubscribe bionews
unsubscribe bioforum
unsubscribe celegans
unsubscribe protista
unsubscribe xtal-log
unsubscribe womenbio
unsubscribe jrnlnote
unsubscribe microbio
end
```

And it turns out to be a happy "ending" after all. Instead of spending time sorting out the wheat Email from the chaff Email, Chris now browses the following corresponding

USENET newsgroups every couple of days, quickly skipping most of the messages except for the ones on topics of interest.

Newsgroup Name	Mailing List Name
bionet.announce	bionews
bionet.general	bioforum
bionet.celegans	celegans
bionet.protista	protista
bionet.xtallography	xtal-log
bionet.women-in-bio	womenbio
bionet.journals.note	jrnlnote
bionet.microbiology	microbio

Chris also discovers, while rereading the BIOSCI instructions, that all of the messages ever posted to the BIOSCI system are archived on the net.bio.net computer. These archives can be accessed through the WWW by opening the URL http://www.bio.net/ or through gopher by opening a gopher session with the host computer net.bio.net. Not only is each newsgroup archived in a separate directory, but searches can also be done on the entire collection of messages to find postings that contain any text of interest. BIOSCI also indexes its Table of Contents newsgroup for biological journals (BIO-JOURNALS by Email, bionet.journals.contents on USENET news), and Chris can pull up references by searching this index. Lastly, BIOSCI maintains an indexed and searchable address database which Chris uses to locate other networked colleagues in the same research field. And so we leave Chris, safe in the knowledge that the newsgroup network has been mastered and is being used productively for research purposes ...

Table 1.

a	Chris Lewis	1	How to become a Usenet site
b	Perry Rovers	1	Anonymous FTP: Frequently Asked Questions (FAQ) List
d	Joel K. Furr	1	DRAFT FAQ: Advertising on Usenet: How To Do It,...Do It
e	Brad Templeton	1	DRAFT FAQ: 10 big myths about copyright explained
f	Mark Moraes	1	Emily Postnews Answers Your Questions on Netiquette
g	Mark Moraes	1	A Primer on How to Work With the Usenet Community
i	Mark Moraes	1	Introduction to news.announce
j	David.W.Wright	1	DRAFT FAQ: Guidelines on Usenet Newsgroup Names
l	Mark Moraes	1	How to Get Information about Networks
o	Mark Moraes	1	Answers to Frequently Asked Questions about Usenet
r	Mark Moraes	1	Rules for posting to Usenet
s	Dave Taylor	1	A Guide to Social Newsgroups and Mailing Lists
t	Mark Moraes	1	Usenet Software: History and Sources
u	Mark Moraes	1	Hints on writing style for Usenet
v	Edward Vielmetti	1	What is Usenet? A second opinion.
w	Mark Moraes	1	What is Usenet?

The "network adventure" described above has undoubtedly happened to many users with some variants. The only slightly optimistic turn in this story is that Chris did not become discouraged when bombarded with Email, but instead reacted positively by learning to use a news program. Unfortunately less motivated people might conclude simply that there was "too much junk" on the net and give up. This would be a sad ending to the tale. The moral to this story is "Use the right tool for the job". For electronic communications with large numbers of people, news software is the right tool. A lot of useful information for those beginning with USENET is contained in the USENET newsgroup news.announce.newusers. Table 1 contains a list of some of the standard messages.

Among the articles above, note the one entitled "Usenet Software: History and Sources" which is very useful for those who do not yet have news software! This is, of course, a "catch-22" since it would obviously be impossible to read the article without news software! Fortunately all of the articles are also accessible using anonymous FTP from `rtfm.mit.edu` in the `/pub/usenet/news.announce.newusers` directory. Reading these articles should be a fine introduction to USENET, particularly

* What_is_Usenet?
* What_is_Usenet?__A_second_opinion.
* Usenet_Software:_History_and_Sources
* How_to_Get_Information_about_Networks
* How_to_become_a_Usenet_site
* Answers_to_Frequently_Asked_Questions_about_Usenet
* Emily_Postnews_Answers_Your_Questions_on_Netiquette
* Hints_on_writing_style_for_Usenet

Unfortunately, it is easy to get wrapped up in browsing the net and reading news to the detriment of one's work. The Internet is in an exciting stage of development and a lot of time is spent on it in "play". A certain level of this is not only healthy, but necessary for the net's development. Used properly, the network can be a tremendous boon to biology research. The best way to ensure that this happens is to have a good local systems and news adminstrator at your institution. Unless you are looking to change careers and wish to drop biology for computer systems adminstration, our advice is to let your systems staff set up news software and other Internet tools for you. Have them get the right tools for you and install them correctly for convenient use. Most news programs have many customizable features. If a program is not working the way you want it to, talk to your admininstrator and see if it can be easily modified. For example, the message storage time or expiration date for each USENET newsgroup is a settable parameter. If this is set too low, your computer might delete messages before you get a chance to read them.

Just as institutions invest in facilities and staff for DNA/peptide sequencing, electron microscopy, etc., so should they invest in easy access to Internet tools. It will be harder to remain competitive in the future without this investment. The Internet is giving a new meaning to the term "scientific community". Electronic newsgroups can quickly focus intellectual power from a variety of disciplines on a research problem and vastly increase the speed of information interchange. It's not difficult to be a part of this exciting technological development, but it is dangerous to ignore it.

From: *Internet for the Molecular Biologist*.
ISBN 1-898486-02-6 ©1996 Horizon Scientific Press, Wymondham, U.K.

10

REAL-TIME COLLABORATION ON THE INTERNET: BIOMOO, THE BIOLOGISTS VIRTUAL MEETING PLACE

Gustavo Glusman, Eric Mercer and Irit Rubin

Introduction

The Internet, the worldwide computer network, is the embodiment of the virtual universe that is usually called "cyberspace". In this virtual space, biologists can find a wide range of tools not generally available in local university or research environments, such as the massive genome-related databases. Even if many of these tools are available locally, one will generally find on the network larger databases and more powerful computers with state-of-the-art software. One of the most valuable resources on the Internet, though, is the diversity of the human beings using it. A large database can be copied, a faster machine can be bought, but there is typically a strong limit to the number of people doing closely related work at a single institution. If there are five people in the world working on a specific system, it is unlikely that they'll get together often to share and discuss their work. In contrast, the Internet provides the connectivity needed to enable these scattered researchers to join efforts. Researchers from many fields now maintain closer connections to their colleagues all over the world via electronic mail and conferences. They don't need to travel physically around the globe to meet each other. Instead, they can 'meet' virtually in cyberspace. The computer network provides a common ground that doesn't have a true physical location, and can be accessed by anyone, from their own networked computer.

For these reasons, considerable work has been put on developing communication protocols and software. There now exists a plethora of computer programs for many different platforms (e.g. UNIX systems, Macintoshes, IBM PC's), each helping users communicate in different ways. These 'Internet services' model well-known, everyday communication systems. For example, sending electronic mail is analogous to sending mail; the UNIX 'talk' protocol is analogous to holding a phone conversation; the World Wide Web (in its simplest form) is analogous to reading magazines, and with its multimedia extensions, it can provide real-time video and audio, like television. Of all the Internet services, the most exciting are those that provide for real-time communication with other people. Such real-time communication in virtual spaces can contribute to the spontaneity and creativity of a scientific discussion: reading announcements from a campus bulletin board can be useful, but sitting in the cafeteria or sprawling on a campus lawn, and brainstorming with your colleagues, can spell the difference between a simple exchange of data and the spirited discussion that inspires new scientific insights. The following section describes a visit to a virtual meeting place for scientists, called BioMOO.

Alice in BioLand

To ease our walk into cyberspace, let us follow Alice, an imaginary biology researcher, as she finds her way to the most powerful services on the Internet, and learns how to benefit from them in her very real research work. Please note that most of the names and situations mentioned in this section are imaginary, and that any resemblance to actual people and happenings are just a happy coincidence. For some time now, Alice's only use of computers has been for handling her lab data and literature searching. She also exchanges sporadic electronic mail with a couple of friends. In fact, she also exchanges regular mail with them, and it doesn't make much of a difference to her. Now Alice heard and read some interesting things about the Internet, so she decided to spend some time investigating... She starts spending more time in the departmental computer room, during incubation times.

Alice starts using Usenet, where she finds many interesting newsgroups, full of messages. She looks at their dates: it's quite amazing! There she is, sitting in front of a computer terminal, reading what people from all over the world wrote yesterday, even this morning! Having a rich imagination, Alice pictures herself standing in a large hall with hundreds of bulletin boards; some empty and some overflowing with messages. She approaches them and reads some messages. Most waste her time, but some are interesting. Walking around the hall, she finds a door labeled 'bionet'. She walks through it and enters another hall, where she finds Biology-related groups. Taking a deep breath, she scribbles a note, describing briefly a technical problem she's been having lately at the bench, attaches it to a bulletin board labeled 'molbio.methds-reagnts', and waits. Nothing happens for a couple of minutes, and Alice's mind wanders a bit...

Her reverie is broken by a wild cry of "He's crazy! He almost killed me!" Somewhat shaken, Alice returns to reality, and notices Robert, another student, sitting in front of a nearby terminal. Alice asks, rather curtly, "Excuse me, but what's your problem exactly?" Flustered, he answers, "Oh, I'm very sorry about that...". Pointing at the terminal he adds, "It's just that he did something very dangerous. I didn't expect him to use that weapon so close to me." Seeing Alice's uncomprehending stare, he explains, "Oh, I see you don't know what I mean. All this is happening in this MUD I'm playing in." She retorts, "This *what*?", but as her lab timer beeps her back to work, she logs out of her terminal and rushes out of the room, somewhat mystified...

The following day Alice goes back to the computer room and remembers the note she posted to the bionet bulletin board, so she walks back into the large hall. She's pleased to see various notes hanging from the one she posted. She reads and finds out that others have had a similar problem, and someone found a solution she can use! Now her curiosity is piqued... how did he think of such a solution? As she writes a new message with many additional questions, she realises she will have to wait - most probably- at least another day to get some answers. If she could only talk to him more directly! "Hello, Alice." Alice looks up, greets Robert, and comments "This bionet thing is really great. I just got an answer to a really serious problem I had." Robert says, "Right, bionet is amazing. I see you're starting to make good use of the Internet." Alice says, "Yes, I had no idea about all this... A pity that it isn't graphical, though... and the low interactivity." Robert says, "Not graphical? Let me show you something, then..." Robert's fingers fly over the keyboard, and suddenly there's a dazzling graphical window on Alice's screen. "This is the World-Wide Web. Using the mouse, you can "move around" and visit linked documents. There are also some very powerful search engines on the web." With a couple of mouse clicks, Robert fetches the query form for a popular search engine, and Alice types "hydra". She explains to him, "You know, we're trying to clone Hydra Hox-like genes..." She clicks on the Submit button and promptly gets a list of 'hits'. Browsing the list, she chuckles, "...no, I didn't mean hydraulic engineering.... Project Hydra? What's that?... Ah, here it is. Loci index for genome Hydra sp. What's GenoBase?" Robert says, "Whoops! I'm going to be late!", leaves Alice browsing the web, and takes a seat in front of the next terminal.

Alice spends some time visiting interesting Biology-related web pages. She finds servers that provide genomic information and methods for analysing it, metabolic charts, image archives, sound files, personal pages with CV's and research descriptions... the wealth of information seems to be inexhaustible! Still, the best she could do to contact the people she "met" through the web, was to send them electronic mail. Some days later, while looking at some lists of "Internet sites for biologists", Alice sees a link that catches her eye:

BioMOO, the biologists' virtual meeting place.
Clicking on it, she gets BioMOO's home page on the web, which is located at `http://bioinfo.weizmann.ac.il/BioMOO`. This page includes a menu:

* BioMOO's purpose
* BioMOO's Frequently Asked Questions
* Visit BioMOO
* Recordings of meetings
* Reference files

Alice clicks on the second item, and starts reading BioMOO's list of frequently asked questions -

```
What is BioMOO?

    BioMOO is a virtual meeting place for biologists, connected to the
Globewide Network Academy. The main physical part of the BioMOO is
located at the BioInformatics Unit of the Weizmann Institute of
Science, Israel.
    BioMOO is a professional community of Biology researchers.  It is
a place to come meet colleagues in Biology studies and related fields
and brainstorm, to hold colloquia and conferences, to explore the
serious side of this new medium.
```

This looks exactly what she was looking for! She resumes reading, and learns how to register in BioMOO and how to learn about other users. She wonders how to reach BioMOO itself :

```
How do I access BioMOO?

    BioMOO's telnet address: bioinfo.weizmann.ac.il 8888
            numeric address: 132.76.55.12 8888
           WWW Home Page: http://bioinfo.weizmann.ac.il/BioMOO
       Multimedia interface: http://bioinfo.weizmann.ac.il:8888

    To connect to BioMOO, you have to:
       telnet bioinfo.weizmann.ac.il 8888
    Once you get the login welcome message, follow the instructions to
connect.
    Note: there are several 'client' programs available, that help you
connect to the MOO more easily. What program to use depends mainly on
what type of computer you're connecting from.
    In parallel, point your favourite WWW browser to one of the URLs
above.
```

Reading further, she also learns what a MOO is: MOO stands for MUD Object Oriented; MUD stands for Multiple User Dimension. A MOO is an object-oriented computer program that allows many users to log in at the same time, and interact among themselves, and with the program. Inside the MOO, everything is represented by objects. Every person, every room, every note, all are represented by objects, that can be looked at, examined and manipulated. She also gets a list of the most basic commands in the MOO - which are fairly intuitive. To talk to other users, you just 'say' what you want. You can 'look' at objects, 'examine' them... The commands look intuitive and English-like, and can be abbreviated to make things easier. In addition, there seems to be plenty of on-line help, and for human help, she can send electronic mail to Gustavo@bioinfo.weizmann.ac.il. Back to BioMOO's home page, Alice clicks on 'Visit BioMOO'. A new window opens on her screen, with BioMOO's welcome message:

```
220- <!- Welcome to...

~%HHH!*%nx.    HM           ~4HHH:    :HHHH~    .xH*"'tx.     .xH*"'tx.
  MMM   'MMM   '"           MMMM:    MXMMX    MMM      #MM.    MMM     #MM.
  MMM    MM"  +nn   .n*%x.  M'MMM  M XMMX   MMM        MMM,   MMM       MMM,
  MMM!**MX.   MM  MM   4Mh  M ?MMM  X~ XMMX  'MMM      MMMX  'MMM      MMMX
  MMM    MMM: MM  MM   'MM  M  ?MMMd"  XMMX  'MMM      MMM'  'MMM      MMM'
  MMM    MMMf MM  MM   'MM  M   MMMf   XMMX   'MMk     :MM#   'MMk     :MM#
..MMMk..HM*"  .MM. 'MX..M*  .:M..  MM  .MMMM.   "Mh...xM"'     "Mh...xM"'


                        ...the virtual meeting place for biologists.

Type:
'purpose'                        to read a statement on BioMOO's purpose
'connect (userid) (password)'    to connect, for example: connect Homer
                                 stu888ph
'guest (name) (any-password)'    to connect as a guest,
'create'                         for information on how to get your own
                                 userid,
'who'                            just to see who's logged in right now,
'@quit'                          to disconnect, either now or later.

Multimedia interface (WWW): http://bioinfo.weizmann.ac.il:8888
For human help please email to Gustavo@bioinformatics.weizmann.ac.il.
220 - - - - - - - - - - - - - - - - - - - - - - - - - - - - - - - - --->
```

Excited, Alice types 'who' to see who's on-line, and sees:

```
Name                     Connected     Idle time    Location
--------                 -----------   ---------    -------------
Maurice (#2365)          an hour       0 seconds    The Lab Wing
Emma (#445)              4 hours       2 seconds    The shore of the lake
Jack [guest] (#155)      3 hours       13 seconds   Natalie's Office
Natalie (#3254)          an hour       22 seconds   Natalie's Office
JasonF (#4663)           an hour       a minute     The shore of the lake
Pamela [guest] (#260)    4 hours       a minute     The Registration Room
Chuck (#1881)            39 minutes    2 minutes    The Central Library
Nassie [guest] (#257)    20 minutes    4 minutes    The Central Room
BioSat1-link (#4046)     9 days        6 minutes    The switchboard
Hans (#554)              2 hours       7 minutes    The shore of the lake
Lion (#3665)             2 hours       8 minutes    The shore of the lake
Armand (#5130)           2 days        a day        Armand's Greenhouse
```

```
Total: 12 users, 8 of whom have been active recently.
```

Alice sees that some of the active users are in the same 'location'... Does this mean they're talking to each other? She types "guest", and gets as answer:

```
Usage:  guest <your-name> <any-password>
```

OK, she thinks, let's try again... She types "guest Alice"....

```
*** Connected ***
There is new news.  Type 'news' to read all news or 'news new' to read
just new news.

The Lounge
A large, silent, dimly illuminated room with lots of people snoring
their real lives away...
A door to the south leads out to the Central Room. (type 'south' to
exit the Lounge; omit the quotes when you enter the command).
A big sign here reads: Type 'tutorial' to learn the basics of MOO, or
if you have a web browser available, type 'web' for a web tutorial.

You see a list of users with bad email address and a Guest Book here.

If you have any problems, try typing:
  page help Can anyone help me?  or  page help <your message here>
```

That did it! She is now in BioMOO's Lounge, and it is quite clear that she has just awakened into this virtual world. Remembering that a guest called Nassie was listed as being in the Central Room, she types "south" as instructed.

```
You move south.

The Central Room
A very large, circular room, its ceiling a transparent dome through
which sunlight streams in.
A large archway leads west into a foyer.
There are doors leading in all directions, labeled with tasteful
signs.
```

```
You see the Research Directory, a public bulletin board, a Jobs/
Postdocs bulletin board, a Lost and Found Box, a road-sign pointing to
the special interests room, a tour dispenser (td), and an arrow
pointing south to the zoo here.
Nassie [guest] is here.
```

```
A sign here says: Check out the new tour dispenser, with "look td."
```

Excited, Alice recalls what she read in the list of Frequently Asked Questions. She types: `"say Hello Nassie!"` and sees:

```
You say, "Hello Nassie!"
```

She waits a bit, but nothing happens. Remembering she can look at anything, she types: `"look at Nassie"`.

```
By definition, guests appear nondescript.
He is awake, but has been staring off into space for 6 minutes.
This guest seems to be connected from 10 from amalia.univis.ac.br.
```

Ah, she thinks... Nassie hasn't been active for some minutes now. He isn't ignoring her, but rather must be doing something else at his computer. Alice decides to wait a bit, and looks at the objects in the Central Room. Wondering what the Research Directory is, she types "`look directory`", and sees:

```
You see a musty, dusty, leather-bound book. You might try something
like 'find neural net in dir'
```

This looks promising! She tries: "`find hydra in dir`", and after some seconds gets:

```
People with 'hydra' matching in their research messages:
Natalia, Bill F, Dr.Jones, Tanya, JasonF and Michelle.
To read someone's research message, type 'research <name>', or 'info
<name>' for more complete information.
```

Aha! JasonF seems to be interested in Hydras, and he's on-line! Alice types "`research JasonF`" and reads:

```
Research interests:
Glycobiology, Carbohydrates, Oligosaccharides, Enzymes, Glycolipids,
Combinatorial Chemistry.
```

Hmm, so searching for just "hydra" was a bit simplistic, as it found also people interested in carbohydrates. On a whim, she checks Tanya's research message:

```
Research interests:
I am currently working on the development of Hydra sp., aiming at
cloning relevant genes.
```

But... that is exactly what Alice is working on! Not believing her luck, she types "`info Tanya`", and learns:

```
>info Tanya
==============================
Tanya (#3466) is in Tanya's Developmental Lab (#3468)
She is sleeping.
Tanya first connected 6 months, 6 days, 13 hours, 6 minutes, and 11
seconds ago.  She last disconnected on Thu Jul 20 03:55:47 1995 IST.

Tanya's real name is Tanya Auerburk and she is a PhD student at the
University of Melbourne.
Her e-mail address is: tanya_auer@unimelb.edu.au

Research interests:
I am currently working on the development of Hydra sp., aiming at
cloning relevant genes.
```

While she's taking note of this valuable information, she notices new activity in her BioMOO window. She reads:

```
Maurice has arrived.
Maurice says, "Hello, Alice. Welcome to BioMOO."
```

That's it! She starts talking -

```
>say Hi there! Do you read this?
You say, "Hi there! Do you read this?"
Maurice says, "Well, sure. First time you use BioMOO? Need help?"
```

She notices this is the first time she communicates with someone else in real time, through the Internet.

```
>say Yes, first time... This is amazing!
You say, "Yes, first time... This is amazing!"
Maurice smiles, "How did you find out about BioMOO?"
>say I found its home page on the web, by chance.
You say, "I found its home page on the web, by chance."
Maurice nods.
>say In fact, this is the first timeMaurice says, "It seems like
you're finding your way around ok. Don't hesitate to ask if you have
any problem."
```

Oops! What was that? It looks like what Maurice said just got mixed with what she was typing... Alice tries again:

```
>say This is the first time I interact in real time over the Internet.
You say, "In fact, this is the first timesay This is the first time I
interact in real time over the Internet."
Maurice says, "Hmmm, I see you can use some help right now. You
connected using a simple telnet program... This is ok for the first
times, but it can be confusing, as you just saw."
Maurice says, "Also, notice you needn't type 'say' all the time. If
you type '"hello', this is the same as typing 'say hello'."
>"testing...
You say, "testing..."
```

```
Maurice says, "Type 'help output' for a workaround to the mixup
problem, but the best solution is to get a special client program."
You say, "What is a client program?"
Maurice says, "It's a specialized program for interacting with the
MOO. Among other things, it separates your input from the output you
get..."
Maurice says, "The best way to get a good client program is to ask
your local system administrator, or some other experienced user. Don't
you know anyone else there that uses MUD's?"
```

Suddenly, Alice understands Robert's weird comments when they first met in the computer room. Noticing he's again working at his terminal, she asks Maurice to hang on, and asks Robert about clients for MUDs. Robert says with a wide grin, "Well, you really are progressing fast with the Internet! So you are playing a MUD now?" Alice answers, "I'm not sure what you mean by 'play' it. It looks like a very serious place to me." Robert asks, "What MUD is that?" Alice says, "BioMOO. A MOO for biologists. I arrived at it through its home page," and shows him BioMOO's address on the web. With a puzzled look, he connects to BioMOO from his terminal, shows up in the Lounge, and goes south to meet Alice in the Central Room. On her screen, she sees:

```
Rabbit [guest] has arrived.
Maurice [to Rabbit]: Welcome. I see you come from Alice's site - can
you help her set up a good client?
Rabbit [guest] says, "Sure, I'm using one now. So this MUD is only for
biologists? I'm quite surprised!"
Maurice says, "What's so surprising? This technology is great for
remote conferencing, so why not use it? :)"
Rabbit [guest] says, "Right... I just hadn't thought about it. Just a
minute, I'll show Alice how to run the client..."
Maurice says, "OK, that's great. In fact, I have to run. Bye Alice,
bye Rabbit, we'll meet again on-line, rest assured!"
Maurice waves.
Maurice goes home.
```

Robert explains that one advantage of MUDs is that there are telnet programs for virtually every different computer type, so anyone can participate without installing special software. Of course, installing such software is recommended anyway... He shows her how to connect to BioMOO using the client he usually uses, and tells her that full explanations on clients can be found on the web at:

http://www.math.okstate.edu/~jds/mudfaq-p2.html,

and that many good clients are available on anonymous FTP from

FTP.math.okstate.edu, in the directory: /pub/muds/clients.

Now comfortably connected with a good client, Alice resumes her exploration of BioLand, the virtual world modeled in BioMOO. Robert (who calls himself 'Rabbit' in the virtual world, as you guessed) helps her find her way around. For example, after wandering into a lab, she feels a bit lost, so he points out that she can always type the command 'map'. She tries it :

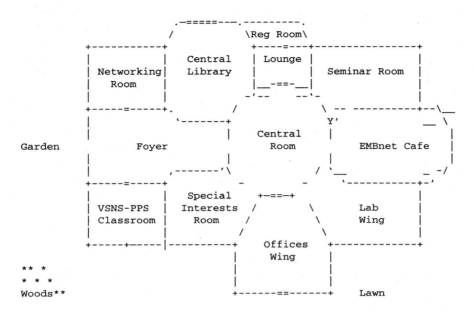

While wondering what lies outside the building, Alice says, "OK, this makes it much clearer... but does this mean I have to find my way through all the rooms if I want to go somewhere? Let's say I want to go to the Central Library." Robert says with a grin, "Well, how about typing 'go to the central library' ?" With an disbelieving smile, Alice tries it, and finds herself teleported to the Central Library, where she sees:

```
    A vast octagonal room whose entire north side is a large paneled
window looking north.  Angled parts of the walls near the ceiling and
floor give the impression of facets.
    The room is dominated by a spiral staircase in its center, leading
down to the generic objects display area, and up to the audiovisual
center.  Doors are set into each wall, but only three are open.
    A door to the south opens to the BioMOO/MOO References room, one to
the west enters the Biology Shelves, and one to the southeast leads
back to the Central Room.
You see a table covered with sheets of paper, a sign, a map, and a
Users directory (ud) here.

Rabbit [guest] materializes out of thin air.
Rabbit [guest] says, "Well, navigation is quite simple, then. You can
'go' to wherever you want, as you did, or 'join' other users. I typed
'join Alice' to come here."

Alice looks at the table...

A rectangular table covered with pages. The nice lettering on them
invites you to read...
    1 - JohnTowell:   Progbit Policy
    2 - JohnTowell:   MOOprogramming FAQ
    3 - Norm:         Journal Clubs on the BioMOO Pamphlet
    4 - El gato:      Note on grammar
    5 - david:        On Objects(OO)
    6 - mouse:        Note for new users
```

Alice says, "Hmm, I see there are various notes on the table, but how can I read them? For example, I'd like to read that note for new users..." Robert says, "Type: `read 6 on table`." Alice asks, "How did you know that was the right command?" Robert says, "You can always `examine` objects, which will give you a list of available commands for them. In this case, examining the table told me that you can: '`read <anything> on table`'." Alice reads the note, which invites new users to peruse the material on the new user's bookshelf, in the reference shelves room to the south. She walks south and finds, indeed, various shelves stock-full with reference material. She takes some notes out of them, reads, and slowly the pieces start falling into place, and she gets a better picture of what this virtual world looks like. "This is like reading a book," she thinks, "except I'm in the book myself. The whole environment is designed to simulate the real world, while simplifying it, so we can concentrate on communication." Alice notices it's time to go, so she mails a note to Tanya, who works - according to the Research Directory - in her same field. That done, she returns to the real world. "In fact," she muses, "while the objects and the place are virtual, the people behind it are real enough. Maybe the distinction shouldn't be between real and virtual world, but rather the former should be called 'physical' world..."

The day after, Alice finds a reply from Tanya in her mailbox, saying she's surprised to learn that someone else is working on her field, and inviting Alice to meet her in BioMOO, giving a time. Alice looks at her watch: the meeting is in about half an hour! The message also says something cryptic about having enough time before the class starts... Alice finds herself again in BioMOO's lounge. She studies the basic tutorial, which can be accessed by just typing "`tutorial`" and following instructions. Suddenly Alice's monitor beeps twice, and she sees:

```
You sense that Tanya is looking for you in Tanya's Developmental Lab.
She pages, "Hello Alice! Come over, join me here in my virtual Lab!
:)"
```

She does and they start talking about the MOO, then about their work, and find they have many interests in common. Their research projects are very related indeed, and they decide to cooperate fully. They both appreciate that both their projects can advance much faster if they share their findings, and now the physical distance is made quite irrelevant, thanks to the Internet. After a while, Tanya suggests that they go to the class. Puzzled, Alice asks what she means, and Tanya explains that a virtual course is going on and they have a study-group meeting. Her curiosity piqued, Alice joins the group and gets a taste of group dynamics in the MOO. For some reason, she had assumed that useful interaction would be restricted to two or three people, but now she sees that the brainstorming can be even more effective with a group of people that work on the same subject!

```
Alice says, "Surely, there must be some limit to the number of people
in one room, just like in the physical world. As fifty people chatting
in one room can make individual conversations difficult, I guess the
text from tens of users can get to be unmanageable?"
JasonF says, "It can, of course. That's why work groups aren't large -
say up to 10 people..."
Emma says, "And there are special types of rooms for handling more
people. For example, the Seminar Room is moderated, which means there
is a speaker queue if needed. Usually, this solution is good for tens
of people at a time."
```

The group meeting continues, and at some point the instructor is at a loss for words and excuses himself while he draws a graph to show what he wants to explain. He then drops a virtual slide, and asks everybody to look at it through the web. Alice "looks" at it by typing the usual 'look' command on the text window, but sees nothing special. She asks the other participants, and they point her at http://www.cco.caltech.edu/~mercer/htmls/ConnectingByWeb.html, where she learns that the WWW can be used as well to view the MOO. Following the instructions and the examples, she points her web client to the address: `http://bioinfo.weizmann.ac.il:8888`, passes the authentication procedure, et voilà! The text-based MOO has also a graphical interface! The graphical interface has various advantages:

- It is more intuitive for new and advanced users - you see a small icon depicting a bulletin board, you click on it, you see a listing of the notes attached to it.
- It makes navigation much simpler - you can click on a graphical map, to reach your destination.
- Iit concentrates many related pieces of information in one page, providing easy customisation of the environments and the objects, with just one form instead of many commands.
- It frees the text window from many activities that don't require full interactivity, making conversation with other users easier to follow.
- One picture is worth a thousand words!

After the meeting, someone suggests that Alice register as a regular user, pointing her at "help registration". She registers, and receives mail from BioMOO with her password and some basic instructions. As a registered user, she can:

- Create her own virtual objects.
- Exchange mail with other registered users, without having to remember Email addresses.
- Transfer texts to her account easily.
- Subscribe to the internal mailing lists, and participate non-anonymously
- Provide information on her research, so other users can find her as she found Tanya.
- Customise her environment and keep her settings between sessions
- ...much more!

A month later, Alice isn't a newbie anymore. She has her own virtual lab in BioMOO, where she keeps a model of her work system; she organises a journal club where a few scientists that work on her narrow field meet periodically; she has set up a colourful poster that summarises her work for other interested people; and is collaborating as a consultant student in an ongoing virtual course!

Strengths and Weaknesses of Internet Communication

Text-based Communication

The main characteristic of BioMOO real-time communication at this time is that it is primarily text-based. It should be noted that intense development of voice communication systems for Internet teleconferencing and MOO communication in particular is currently underway. Most of these, however, will require that the computers of the participants be directly connected to the Internet, which is not generally the case at this time. We can probably expect to see such systems tested in BioMOO by January 1996, but since text-based communication is the primary means at our disposal now, we will limit our discussion to that format. Also, it is worth mentioning that the most common language of communication in BioMOO is English, although when non-English speakers meet there, they typically use whatever languages they share that can be displayed using the general ASCII character set.

Text-based communication has the advantages of being typically easier for non-native speakers to follow (text is available for re-reading on the screen), free of accents (though not necessarily free of confusing grammatical or spelling errors), and being simple to record and edit. It has the disadvantage of requiring typing skills, and typically being much slower than speech. Because of its slowness, many users will actually perform other tasks concurrently during a text-based communication session. The windowing systems on most computers today allow one to do word-processing or on-line literature searches while participating in a text-based conversation. A common method to increase the efficiency of presenting information, at least in formal presentations, is to prepare much of the presentation text beforehand and either use a windowing system to paste the text into the BioMOO speech system, or record it on a specialized BioMOO object that allows selected portions of the text to be presented and appear to listeners as if it was spoken by the presenter.

Dispersed Participants and Timezones

BioMOO users live all over the world, thus it is not possible to find a time when all are likely to be awake and available. Presently, however, most BioMOO users live in North America and Europe. Meeting times between 16:00 and 19:00 GMT are most reasonable for most BioMOO users. This corresponds to 8 AM to 11 PM for the west coast of North America, 5 PM to 8 PM for western Europe, and 6 PM to 9 PM for eastern Europe and the Middle East. Note that this generally precludes participation of users in India, Australia, and the Far East.

Graphical Data Presentation

One of the main differences between the different communication formats, that will be discussed in the next section, is how much each depends on accessing image data over the Internet. For instance, seminars generally require the presentation of "slides." Images can be posted to a publically accessible site and retrieved by participants either by gopher or FTP before the presentation (if the site supports those access methods), or during the presentation using the World Wide Web and an appropriate web browsing program. Unfortunately, not all BioMOO users have access to a graphical web browser. For this reason, the most effective way to present such images appears to be by putting them on an FTP or gopher server well before the presentation, announcing their availability and location, and then providing appropriate links to them during the presentation for web access by those with that capability. The long term solution is for more BioMOO users to get graphical web access, and while this is clearly the trend, it appears no more than one third to one half of BioMOO users could use this system as of February 1995.

Communication Formats

Journal Club

This is the presentation format that has been used most often in BioMOO. Note that the name for this format varies, but it generally consists of the presentation of one (or two related) papers, including an introduction to the field, presentation of the results, and discussion among all participants. It has the advantage of not requiring participants to obtain images over the Internet, since papers presented are usually in widely available journals. In addition, it is probably the most discussion-provoking of the formats, taking advantage of BioMOO's main strength: providing a channel for real-time, dispersed participant, international communication. Its main disadvantages are that it doesn't involve the presentation of new information (which is generally most interesting to researchers), and it requires all participants to be present at the same time. Also, the usual slowness of text communication means that what would be a thirty minute presentation and discussion using speech, typically takes an hour by text. Of course, the participants didn't have to spend hours or days reaching the meeting place.

Poster Session

Not all formal communication methods require synchronous participation by all users. For instance, a poster session can be visited at any time. Posters can be made available for access at people's convenience and then have associated discussion sessions scheduled where the poster presenters are available in BioMOO for questions. This powerful combination allows people to examine the presented data, and then meet the presenter for relatively informal discussion. The ability to present posters integrating multimedia data and accessible via graphical web browsers makes this probably the most exciting recent development in scientific communication. The addition of BioMOO-based discussion to such a presentation brings together all the best aspects of a live poster session with internet-based real-time communication systems. The only significant disadvantage is that poster sessions are generally likely to work best when most of the participants have access to graphical web browsing programs. Well written figure text, however, can provide a great deal of useful information for non-graphics users. BioMOO has had a small poster session successfully running for months. Specialized objects have been provided to simplify construction of posters, and provide a standard presentation style to increase the rate participants become accustomed to the poster format. BioMOO provides (space-limited) publically accessible web sites for BioMOO members to post data presented in poster sessions.

Interview

This format generally consists of one person interviewing a guest speaker, who may also present a short talk with or without graphical data presented. One trial in this format has been presented in BioMOO, and was quite successful. It opened with a short set of questions and answers between the interviewer and guest, and was followed by an open question and discussion session with audience participation. This presentation system is useful when the goal is to provide communication and discussion covering a broader set of information than is usually presented in a seminar. Topics for discussion can be selected by the interviewer, guest, and audience, and discussion proceeds to whatever depth and extent the participants wish. It has the disadvantages of requiring all participants to be present together, a slow communication rate, and typically a more shallow exploration of any single topic. It seems likely to be a very effective way to present an interesting person with expertise in some field of study to an audience less familiar with that field, in a relatively informal setting.

Seminar

At this time, no graphical seminars presenting original data have been held in BioMOO. The primary reason for this has been the limited number of people using graphical web browsers, and the only recent development of BioMOO's web interface. The former is likely to be the main disadvantage of presentations in this format in the near future. However, the potential for using this well-established system for conveying recent scientific work makes it likely that trials will soon begin. Together with the poster session, these comprise the primary formal presentation methods used in contemporary scientific meetings. The development of Internet-based systems that offer these presentation methods is likely to revolutionize communication between scientists.

Study group

During the last months, two graduate-level courses have been held in BioMOO. Typically, the study material was made available to the students in hypertext or PostScript form on the World-Wide Web. The students browsed it at their own pace, and met regularly with their instructor and the other students in their study group to discuss the weekly chapter, raise questions, and discuss exercises. These meetings were held in BioMOO, in a virtual classroom where helpful objects were made available. The study groups were composed of the instructor and up to five students; the discussion was unmoderated and free-flowing. This format proved to be effective for student-teacher and student-student interaction. The participants were able to behave and communicate as they would have done in a physical meeting: the hallmark of successful network interaction.

Spontaneous interaction

In addition to the organised communication formats described above, BioMOO's strength resides in unplanned, spontaneous interactions between its users. In Alice's story, she meets Maurice by chance. Such serendipitous meetings happen all the time, without anyone planning them, and they frequently result in useful exchanges between the participants. At another level, Alice contacts Tanya after learning about her from the Research Directory, and their meeting takes place in an informal setting. These two interaction modes are analogous to meetings between researchers during and after "real life" conferences, which are propicious opportunities for exchanging ideas and initiating collaboration. In this sense, BioMOO can be regarded as an ongoing conference, its topic evolving with the interests of its registrants.

Conclusions

BioMOO and other text-based communication systems are clearly still a developing tool for scientific communication. While it seems certain that seamless speech and live video-based systems will eventually be typical of international scientific communication, many people are finding that the more limited systems now available can provide an exciting and viable means for enhancing their professional interactions. In addition, as an object-based virtual reality (VR) system, BioMOO provides what is, for many, the first taste of what is widely anticipated to be the VR experience of the future: full sensation immersive VR. It seems likely that nearly all of the sorts of communication tools and virtual objects that are being developed in text-based VR systems will be applied to immersive VR, and it in fact provides an outstanding proving ground for such development. The hardware obsession of present VR researchers shouldn't obscure the fact that software is what finally makes a system useful. Discovering the tools that make a VR system useful, and developing the parameters and features of such tools, is a vital role for text-based VR systems. Beyond its role as a development site, however, BioMOO has already proven to be a useful tool for scientific collaboration. The observation that scientists using BioMOO interact essentially as they do in person, makes it clear that this medium can serve as a valuable tool for enhancing long-distance, real-time professional communication.

Acknowledgements

We wish to thank our BioMOO co-administrator John Towell for his helpful comments and suggestions, and BioMOO's system administrator, Jaime Prilusky, for making it all possible.

Further reading

Anderson, C. 1994. Cyberspace offers chance to do 'virtually' real science. Science. 264: 900-901.

Bachrach, S. M. Electronic conferencing on the Internet: The first Electronic Computational Chemistry Conference. J. Chem. Info. Computer Sci. (In press).

Curtis, P. 1992. Mudding: Social Phenomena in Text-Based Virtual Realities. In: Proceedings of the 1992 Conference on Directions and Implications of Advanced Computing, Berkeley, May 1992. Also available as Xerox PARC technical report

CSL-92-4 or at `ftp://parcftp.xerox.com/pub/MOO/papers/` `DIAC92.ps`

Curtis, P. and Nichols D. A. 1993. MUDs Grow Up: Social Virtual Reality in the Real World. Available at `ftp://parcftp.xerox.com/pub/MOO/papers/` `MUDsGrowUp.ps`

Evard, R. 1994. Collaborative networked communication: MUDs as systems tools. In: Proceedings of the Seventh Systems Administration Conference (LISA VII). Available http: Hostname: `www.ccs.neu.edu`   Path: `/home/remy/` `documents/cncmast.html`

Hardy, B. J. 1995. Tales from the launching of global electronic conferencing in computational chemistry. Chemical Design Automation News. (In press).

Krieger, J.H., and Illman, D.L. 1994. Internet offers alternative ways for chemists to hold conferences. Chem. Engi. News. December: 29-40.

Thornley, B. 1994. Travels on the net. Technology Rev. July: pp. 20-31.

Towell, E. R., Yim, A. and Lam, T. 1995. Utilization of Internet Services and the Teaching of Internet in Business School. J. Computer Info. Systems. (In press).

Towell, J. F., Hansen, P., Mercer, E., Leach, M., Rubin, I., Prilusky, J. and Glusman, G. 1995. Networked virtual environments and electronic conferencing. In: Proceedings of the First Electronic Computational Chemistry Conference [CD-ROM]. S.M. Bachrach (Ed.). ARInternet. (In press).

Towell, J. F. and Towell, E. 1995. Internet Conferencing with Networked Virtual Environments. Internet Research . 5: (In press).

Transcripts of some Journal Clubs held in BioMOO can be accessed at: `http://bioinfo.weizmann.ac.il/BioMOO/Meetings`

From: *Internet for the Molecular Biologist.*
ISBN 1-898486-02-6 ©1996 Horizon Scientific Press, Wymondham, U.K.

11

INTERNET RESOURCES FOR HUMAN AND MOUSE MOLECULAR GENETICS

M. A. Kennedy

Introduction

It is now over five years since Walter Gilbert, in a prescient commentary in "Science", urged molecular biologis ts to develop computer literacy and to "hook our individual computers into the worldwide network that gives us access to daily changes in the databases and also makes immediate our communications with each other" (Gilbert, 1991). If you have not yet heeded this advice then you have much to gain by learning now because, as Gilbert anticipated, the Internet is where we find many of the key tools with which we perform our craft. This chapter introduces some of the Internet resources that are of use to researchers working on human and mouse genetics. Given the breadth of this field, and the rate at which Internet resources grow and change, this cannot be a comprehensive review; the aim is simply to give sufficient information to get new users familiar with key resources. The main resources described here offer on-line guides and help documentation that are updated with each improvement to the service offered, and it is neither possible nor wise to commit anything other than an introductory overview to paper! Once a basic familiarity is achieved, it is straightforward to explore the resources in depth and extend your skills without recourse to a written manual.

Most of the services described in this chapter can be accessed in several ways, and where possible details for Email servers, gopher clients, or WWW browsers are given. Some of the databases also offer FTP access to provide transfer of large amounts of information to your own computer. Although in most cases similar information can be obtained by several routes, WWW sites for browsers such as Mosaic or Netscape are quickly becoming the standard for most Internet resources.

Newsgroups

The best way to keep aware of the services that are available is to monitor appropriate Usenet newsgroups, particularly the BIOSCI/Bionet groups (Bleasby *et al.*, 1993). The FAQ (frequently asked questions) file can be obtained by Email from `biosci@net.bio.net`, or viewed at `http://www.bio.net/BIOSCI/ biosci.FAQ.html`. Most new services and modifications to existing services are announced in these groups, and they are an appropriate forum for any questions you may have in specific areas. Checking the FAQ file or searching the Bionet archives is often a good way to rapidly obtain information on subjects already discussed in any of the Bionet groups, and in any event this should be done before posting to the newsgroup. Searches of the Bionet News archives can be performed by:

Email server	Email message containing your text query to `waismail@net.bio.net`.
gopher	`net.bio.net` or `iubio.bio.indiana.edu`)
WWW	`http://www.bio.net/`
	`http://iubio.bio.indiana.edu/`).

The following small subset of BIOSCI/Bionet groups is likely to be of general interest to human and mouse molecular genetics researchers:

bionet.diagnostics	Discussions of problems and techniques in all fields of diagnostics
bionet.diagnostics.prenatal	Discussions of problems and techniques in prenatal diagnostics
bionet.genome.chromosomes	Discussions about mapping and sequencing of eucaryote chromosomes
bionet.molbio.gdb	Messages to and from the Genome Data Bank staff
bionet.molbio.gene-linkage	Newsgroup for genetic linkage analysis
bionet.molbio.genome-program	NIH-sponsored newsgroup on human genome issues
bionet.molbio.methds-reagnts	Requests for information and lab reagents, discussion of molecular biology methods.
bionet.software	Information on software for the biological sciences
bionet.software.www	Announcements about new information resources in biology which can be accessed via the WWW.

Mailing Lists

There are several relevant mailing lists which allow interaction with colleagues who have similar interests. Each message posted to the list-server is distributed to all other list subscribers. With mailing lists, messages arrive in your electronic mailbox, and you can choose to read or delete them. This is a particularly useful option when only a low frequency of messages is expected, such as in some specialist subject areas, or if access to the information needs to be regulated or monitored. For the mailing lists below, the name of the list and the address to which subscription and help requests should be sent are listed. Sending a message containing only the word "help" (in the subject line or body of the message) to that address normally generates an automatic, Email response containing more detailed information about the list. The body of a subscription request should normally conform to the format:

subscribe list_name your_first_name your_last_name.

CCR-Announce-Request (Coriell Cell Repositories)

Subscription   `ccr-announce-request@arginine.umdnj.edu`

Description   This list is maintained by the Coriell Cell Repositories service, at the NIH National Institute of General Medical Sciences, to inform the research community about the availability of new mapping panels, new cell culture and DNA collections, and other relevant topics. To subscribe, send Email to the above address with the word "subscribe" in the subject field. To unsubscribe or obtain help, include the word "unsubscribe" or "help" in the subject line of your message.

FAMCAN-L (Familial Cancer)

Subscription   `listproc@moose.uvm.edu`

Description   This list supports discussion among health care professionals and academic researchers on issues related to the identification, care, and management of patients and families with cancer predisposition syndromes. A broad range of topics are covered, including all the familial cancer syndromes, counselling, molecular diagnosis, ethical issues, patient management, disease surveillance.

HUM-MOLGEN (Human Molecular Genetics)

Subscription   `LISTSERV@NIC.SURFNET.NL`

Description   A new and fast-growing list dealing with all issues relating to human molecular genetics. The list is moderated by several editors, and it is subdivided into sections. You can choose which subject areas you wish to receive mail about (such as literature, biotechnology/molecular biology, ethical issues, diagnostics, collaborations and so on) by sending an Email command to the list-server.

This list also has an associated WWW site (see below) through which you can subscribe to HUM-MOLGEN and set the topics you wish to receive, as well as retrieving information and linking to relevant Internet services.

Mx-DIAG-L (Molecular Diagnostics)

Subscription   `LISTSERV@ALBNYDH2.bitnet`

Description   Unmoderated list covering molecular diagnostic applications in pathology and diagnostics, including technical discussions, regulatory information, meeting announcements, managerial issues, and other relevant topics.

MGI-LIST, MGD-LIST and Others from Jackson Laboratory

Subscription   `listserver@informatics.jax.org`

Description   The Jackson Laboratory (Bar Harbor, Maine) is a key resource for information on the laboratory mouse. At present, 26 electronic mailing lists are distributed by the Laboratory. All lists are subscribed to by inserting the name of the list in the standard subscription message, as detailed above.

MGI-LIST deals with topics concerning bioinformatics and the mouse genome. It provides a communication forum for any researchers with an interest in the mouse genome. MGD-LIST is a help and communcation forum for users of the Mouse Genome Database (see below). Some of the other lists maintained at the Jackson Laboratory are chair-list (names and contact addresses for chair of each mouse chromosome committee); nomen-list (mouse genome nomenclature announcements and discussion); and lists for each mouse chromosome. There are 22 chromosome lists named according to the format chr#-list, where the # is replaced by 1-20,X or Y.

Rodent-Research

Subscription   `Majordomo@cco.caltech.edu`

Description   This unmoderated list is for announcements, questions, and discussion related to the use of rodents in biomedical research. Message topics range from methodology through research guidelines, embryonic stem cell culture, and many other relevant subjects. Appropriate forum for announcements of particular interest to any researchers using rodents in their work.

Databases

It is fair to say that Internet resources have in large part replaced the reference textbooks of our predecessors. The evolution of these on-line reference resources was necessitated by the sheer bulk of molecular biology data, the speed with which it accumulates, and the need to combine or compare these data from several sources. The success of such resources was stimulated by the development of tools that allow virtually instantaneous access from anywhere in the world. With the increasing sophistication of Internet tools like the WWW, there is now a progression towards a "federation" of databases where hypertext links are used to draw information on related topics together. IGD and several other database browsers listed below use this approach, and several discrete databases such as GDB, OMIM and MGD are developing this functionality. For example, searching OMIM by WWW brings back a disease entry that has hypertext links to relevant DNA sequence or protein database entries.

Database Browsers

The following resources are browsers that manage database access. The browser software in each case facilitates either integrated analysis or menu-based interrogation of several databases, making these the preferred methods for executing comprehensive database searches.

BCM Search Launcher

WWW   `http://gc.bcm.tmc.edu:8088/search-launcher/`
             `launcher.html`
Description   A well organized site that provides a single point of entry for searching many databases. Includes a multiple sequence alignment service, and several useful human genome resources.

DBGET

WWW   `http://www.genome.ad.jp/dbget/dbget.html`
Gopher   `gopher.genome.ad.jp`
Email   `dbget@genome.ad.jp`
Description   DBGET is a simple database retrieval system that permits searching and extraction of entries from many databases. DBGET can be used by Email, gopher, and WWW. The DBGET service via WWW allows retrieval of related entries in different databases by hypertext links.

ExPASy Molecular Biology Server

WWW   `http://expasy.hcuge.ch/`
Anonymous FTP   `ftp://expasy.hcuge.ch/`
Description   The ExPASy WWW server from Geneva University (Appel *et al.*, 1994) allows browsing of a number of databases including SWISS-PROT and PROSITE as well as other cross-referenced databases (such as EMBL/GenBank, OMIM, Medline), and it provides access to various related documents and sequence analysis tools.

GENOBASE

WWW   `http://specter.dcrt.nih.gov:8004/`
Description   The NIH GenoBase WWW Server incorporates and links the contents of several large nucleic acid and protein databases, including EMBL and SWISS-PROT. The interface allows you to view information for a given GenoBase object (an EMBL locus, a SWISS-PROT peptide, a PROSITE pattern etc.), and to easily move from object to object within GenoBase.

IGD (Integrated Genomic Database)

WWW   `http://genome.dkfz-heidelberg.de:80/igd-docs/`
Email   `Genome@DKFZ-Heidelberg.de`
Anonymous FTP `genome.dkfz-heidelberg.de`
Description   The Integrated Genomic Database (IGD) is an information management system for human genome researchers which interconnects existing molecular biology databases and analysis tools (Ritter, 1995). Databases which are interconnected in IGD include genome map and resource databases from CEPH-Genethon and CHLC, sequence databases such as EMBL, GENBANK, PIR, and SWISS-PROT, and the genetics databases GDB and OMIM. There is a FAQ guide for IGD at: `http://genome.dkfz-heidelberg.de:80/igd-docs/faq.html`.

OWL

WWW   `http://www.gdb.org/Dan/proteins/owl.html`
Description   OWL is a non-redundant protein sequence database which ties a number of databases together (Bleasby and Wootton, 1990). The Web version derived from OWL has hot links to all of the major nucleic acid and protein sequence databases, and to other databases including PDB (The Protein Databank, for 3D structures), The EC Enzyme Classification Database, PROSITE, OMIM, and REBASE.

SRS (Sequence Retrieval System)

WWW   `http://www.ebi.ac.uk/srs/srsc/`
Description   A comprehensive browser for molecular biology databases ranging from those for sequences and structures through to transcription factors and literature (Etzold and Argos, 1993).

Notes: Several other sites offer the SRS, including the Swiss EMBnet node (`http://www.ch.embnet.org/srs/srsc`) and the Weissmann Institute (`http://dapsas.weizmann.ac.il/bcd/srs/srsc`).

WWW Entrez

WWW   `http://www.ncbi.nlm.nih.gov/Search/index.html`
Description   The ENTREZ network browser, provided by the National Center for Biotechnology Information, allows searching and browsing of protein and nucleic acid databases, and related references from MEDLINE. For full functionality, Institutions should register to use Network Entrez (with a requirement to install and maintain software on a local machine) by sending a message to `net-info@ncbi.nlm.nih.gov`.

Major Sequence Databases

The major nucleotide and protein sequence databases share information so that their content is equivalent. In almost all cases, a sequence you require from Genbank will also be available from EMBL or DDJB. Various forms and subsets of these databases are usually available by anonymous FTP, but be warned - for the major databases these can be enormous files. Methods for searching and retrieving sequences are detailed in chapter 7, and minimal reference is made here to these methods. The DNA and protein sequence databases, in addition to their primary function as repositories of molecular sequence data, are also a rich source of molecular biology information and references. Keyword searches of the databases by gopher, WWW or Email server as outlined below, can often yield useful information not easily obtained by other means. When seeking information on a particular genetics subject it is worth considering, in addition to your CD-ROM based literature search, an Internet keyword search of the sequence databases for references (often unpublished), addresses, and other relevant data.

DBEST

WWW   `http://www.ncbi.nlm.nih.gov/dbEST/index.html`
Gopher   `gopher.gdb.org`; go to the "Sequence Databases..." directory.
Description dbEST is a division of GenBank for cDNA sequence and mapping data, particularly for single-pass sequencing projects that generate large numbers of expressed sequence tags (Boguski *et al.*, 1993). The sequences in dbEST are incorporated in GenBank, but annotation in dbEST is more comprehensive and includes contact information, genetic map locations (when available), and instructions on obtaining physical DNA clones from the American Type Culture Collection and other sources.

DDJB

WWW   `http://www.nig.ac.jp/`
Gopher   `gopher.nig.ac.jp`
Anonymous FTP `ftp.nig.ac.jp`
Description   The DNA Data Bank of Japan was established in 1986 and operates in a similar manner to EMBL, Genbank. and GSDB.

Notes: More information about DDJB can be found at:`http://www.nig.ac.jp/intro-e.html#abriefintor`. Gopher searches can be performed and Email servers for searching DDJB can be accessed at `fasta@nig.ac.jp` or `blast@nig.ac.jp`. For more details, send a message to these addresses with the word "help" in the body of the mail message.

EMBL

WWW	`http://www.ebi.ac.uk/ebi_docs/embl_db/ebi/topembl.html`
Gopher	`gopher.ebi.ac.uk`
Anonymous FTP	`ftp.ebi.ac.uk`
Description	The EMBL Nucleotide Sequence Database (Higgins *et al.*, 1992) is a comprehensive database of DNA and RNA sequences collected from many sources or submitted by researchers and maintained at the European Bioinformatics Institute (EBI - see below), Hinxton Hall, Cambridge (UK). Sequence submission can be by WWW, by Email form or by Authorin (see Genbank section).

Notes: Full details for submission to this database are given at: `http://www.ebi.ac.uk/ebi_docs/embl_db/ebi/authorinfo.html`. Sequence retrieval can be by:

SRS (see above)	`http://www.ebi.ac.uk/srs/srsc`
WWW	`http://www.ebi.ac.uk/htbin/emblfetch`
Email	`Netserv@EBI.AC.UK` (send a message containing help)

Several homology search servers are also available, details can be found at: `http://www.ebi.ac.uk/ebi_docs/embl_db/relnotes42/network_services.html#Email`. Information about the server for FASTA, a commonly used search program, can be obtained by sending a message containing the word "help" to `Fasta@EBI.AC.UK`.

GENBANK

WWW	`http://www.ncbi.nlm.nih.gov/Genbank/index.html`
Gopher	`gopher.nih.gov`; go to the "Molecular Biology Databases" directory
Anonymous FTP	`ncbi.nlm.nih.gov`
Description	GenBank is the NIH genetic sequence database, a collection of all known DNA sequences (Benson *et al.* 1994).

Notes: A full description of Genbank is available at `http:www.ncbi.nlm.nih.gov/Genbank/nar.edit.html`. Submission of sequences can be made by:

BankIt (a WWW form)	`http://www.ncbi.nlm.nih.gov/BankIt index.html`
Authorin (a PC or Mac program)	`http://www.ncbi.nlm.nih.gov/Genbank authorin.html`
Email	`authorin@ncbi.nlm.nih.gov`.

Completed entries can be mailed either electronically or by post to NCBI. For information about accessing Genbank by Email server, send a message containing the word "help" to either `blast@ncbi.nlm.nih.gov` (for homology searches) or `retrieve@ncbi.nlm.nih.gov` (for text searches and sequence retrieval).

GSDB

WWW   `http://www.ncgr.org/gsdb/gsdb.html`
Email   `gsdb@gsdb.ncgr.org`
Description   A DNA sequence repository at the National Center for Genome Resources in Santa Fe, New Mexico, that cooperates with the other major databases and that can handle the needs of major genome sequencing laboratories.

Notes: A full description of GSDB is given at: `http://www.ncgr.org/gsdb/aboutgsdb.html`. For information on sequence submission, send a message containing the word "help" to `datasubs@gsdb.ncgr.org`. Alternatively access the WWW page at: `http://www.ncgr.org/gsdb/data_ret.html`. The GSDB Web server (`http://www.ncgr.org/gsdb`) provides several different access mechanisms.

PIR

WWW   `http://www.gdb.org/Dan/proteins/pir.html` (PIR Web server)
Gopher   `gopher.nih.gov`; go to the "Molecular Biology Databases" directory.
Description   PIR is a protein sequence database sponsored by the National Biomedical Research Foundation (US) that links a number of protein and nucleic acid sequence databases and allows analysis of these (Sidman *et al.*, 1988). The WWW version offers hot links to the major nucleotide databases, the EC Enzyme Classification Database, Genome Database (GDB) and to literature references.

SWISS-PROT

WWW   `http://www.ebi.ac.uk/ebi_docs/swiss-prot_db/swisshome.html`
Gopher   `gopher.nih.gov`; go to the "Molecular Biology Databases" directory.
Description   A database of protein sequences derived from translations of DNA sequences in the EMBL Nucleotide Sequence Database, adapted from PIR (Protein Identification Resource), extracted from the literature and submitted by researchers (Bairoch and Boeckmann, 1991). It features high quality annotation and cross referencing to several other databases.

Genes, Markers and Disease Databases

CAD (Chromosome Abnormality Database)

WWW   `http://www.hgmp.mrc.ac.uk/local-data/Cad_Preamble.html`
Gopher   `gopher.hgmp.mrc.ac.uk`; go to "HGMP Databases/" directory.

Description   The Chromosome Abnormality Database (CAD) stores records of
                   human chromosome abnormalities, and can be freely searched.
                   Established in early 1991, the CAD contains records of acquired
                   and constitutional chromosome abnormalities from cytogenetics
                   laboratories throughout the United Kingdom. The database includes
                   information on the availability of cell lines and other stored material.

Encyclopedia of the Mouse Genome (by Gopher)

Gopher   `gopher.gdb.org`; go to the "Search databases " directory
Description   The gopher version of The Encyclopedia of the Mouse Genome
                   allows you to view the genetic maps of each mouse chromosome
                   (if your client supports graphics), search for specific loci on these
                   maps as well as search for references and notes associated with
                   each locus on the map. Information about the fully implemented
                   version of the Encyclopedia, which requires locally mounted
                   software,   can   be   found   at: `http://`
                   `www.informatics.jax.org/encyclo.html`.

GDB (Genome Database)

WWW   `http://gdbwww.gdb.org/gdbhome.html`
Gopher   `gopher.gdb.org`; in the "Search Databases at Hopkins"
                   directory
Email   `mailserv@gdb.org`
Anonymous FTP   `ftp://ftp.gdb.org/`
Description   A comprehensive, fully searchable repository of human gene
                   mapping information, cross referenced to other relevant databases
                   including OMIM and the Genome Sequence Data Bank. GDB stores
                   information on the locations of genes and polymorphic DNA
                   markers, the reagents used to identify and locate them, and
                   associated references and contact addresses (Fasman *et al.*, 1994).

Notes: A general overview of OMIM/GDB is located at; `http://`
`gdbwww.gdb.org/intro.html#overview`. An on-line tutorial for GDB is
available at;`http://www.caos.kun.nl/genomics/GDB_OMIM_TOC.html`.
There are mirror sites for the GDB Web browser around the world, a full listing of
which can be found at `http://gdbwww.gdb.org/mirror_sites.html`.
General enquiries about accessing or using GDB can be directed to `help@gdb.org`.
Fast and simple text-based searches of GDB can be performed by Email (send your
query or a message containing the word "help" to the above Email address) or by using
the WWW server at: `http://gdbwww.gdb.org/wais/gdb-wais.html`.
         However, the GDB WWW Browser is better suited for complex searches,
as it provides fill-in forms for each type of primary data object in the database (loci,
probes, maps, polymorphisms, mutations, cell lines, recombinant libraries, citations,

and contact addresses for editors and contributors). This can be accessed at `http://gdbwww.gdb.org/gdbdoc/topq.html`. Entries matching query statements are returned from the database with hotlinks that allow easy location and tracing of related information. A forms-capable WWW client program is required for using this browser properly.

MGD (Mouse Genome Database)

WWW	`http://www.informatics.jax.org/mgd.html`
Gopher	`hobbes.informatics.jax.org`; go to the "Information at The Jackson Laboratory/Informatics Software and Databases" directory
Email	`mgi-help@informatics.jx.org`
Description	MGD (Nadeau *et al.*, 1995) is the mouse equivalent of GDB. It provides a repository of experimental genetic data derived from research on the laboratory mouse, including information on mouse loci, homology between mouse and many other mammalian species, polymorphisms, probes, clones, PCR primers and mapping data, along with full references for all entries.

MGD is available by gopher or via WWW for either forms-capable browsers at; `http://www.informatics.jax.org/doc/frmsupport.html` or text-only browsers `http://www.informatics.jax.org/noforms.html`. Results are returned with hypertext links to related information in the database. The Email list MGD-LIST offers support for users of MGD (see above). General enquiries about MGD can be directed to `mgi-help@informatics.jax.org`.

OMIM (Online Mendelian Inheritance in Man)

WWW	`http://gdbwww.gdb.org/gdbhome.html`
Gopher	`gopher.gdb.org`; in the "Search Databases at Hopkins" directory.
Email	`mailserv@gdb.org`
Description	Subtitled "A catalog of human genes and genetic disorders", this is the electronic form of Victor McKusick's extraordinary compilation, Mendelian Inheritance in Man (McKusick, 1994; Pearson *et al.*, 1994). It contains a wealth of data on human genetic traits and inheritance patterns, and each disease entry consists of a comprehensive literature review with full cross referencing to other relevant databases. Other data in OMIM include the "MIM Number" or unique reference for each condition, details of allelic variants, and clinical synopses for each condition. OMIM can be entered from the Genome Data Base (see below), and conversely GDB can be entered from OMIM.

OMIM is an exceptionally good way of finding out how much is known about a disease without leaving your desk. OMIM text entries are comprehensive, giving chronological overviews and helpful summaries of the literature with extensive reference lists. The powerful search functions, cross referencing (by hypertext links in the WWW version), and continual updating of this database make it an indispensible tool for anyone interested in human disease.

Notes: Help with search functions for OMIM is available on WWW at: `http://gdbwww.gdb.org/omimdoc/query_help.html` and an on-line tutorial is available at: `http://www.caos.kun.nl/genomics/GDB_OMIM_TOC.html`. A general overview of OMIM/GDB is located at: `http://gdbwww.gdb.org/intro.html#overview`. Users with simple queries may search GDB/OMIM via Email. To get instructions on querying the Email server, send a message with the word "help" in the body of the message. General enquiries about accessing or using OMIM can be directed to `help@gdb.org`.

TBASE (Transgenic Animals)

WWW	`http://www.gdb.org/bio/search-sf2/TBASE/tbase`
Gopher	`merlot.welch.jhu.edu`; go to the "Search databases at Welchlab" directory.
Description	TBASE (transgenic animal/targeted mutation database) is a searchable database that stores information about specific transgenic animals, including transgenic and knockout mice (Woychik *et al.*, 1993). Each entry gives a comprehensive summary of an individual mutant line, including details of the derivation methods, genetic background, phenotype and other relevant experimental data, and addresses of authors and contacts.

Tumor Gene Database

WWW	`http://kiwi.imgen.bcm.tmc.edu:8088/bio/bio_home.htm`
Gopher	`condor.bcm.tmc.edu`
Description	This is a database of genes associated with tumorigenesis and cellular transformation, maintained at the Baylor College of Medicine, Texas. It includes oncogenes, proto-oncogenes, tumor supressor genes, tumor-associated chromosomal breakpoints and viral integration sites, and other relevant genes and chromosomal regions. Entries contain extensive data on such things as chromosomal location, properties, gene products and literature references.

Other Useful Databases

EPD (Eukaryotic Promoter Database)

WWW	`gopher://gopher.gdb.org/77/.INDEX/epd`
Gopher	`gopher.gdb.org`; go to the " Search Databases at Hopkins..." directory.
Anonymous FTP	`ncbi.nlm.nih.gov`; in the /repository/EPD/db/ directory.
Description	EPD (Bucher and Trifonov, 1986) provides information about eukaryotic promoters available (but not necessarily annotated as

such) in the EMBL Nucleotide Sequence Database, with the aim of assisting analysis of eukaryotic transcription signals. A copy for local use can be retrieved by anonymous FTP, or keyword searching can be done by gopher.

Molecular Biology Vector Sequence Database

WWW	`http://biology.queensu.ca/~miseners/vector.html`
Description	A growing compilation of vector sequences. Particularly good for contemporary cloning vectors, as this site provides links to those companies that keep databases of their own vector sequences.

ProDom

WWW	`http://protein.toulouse.inra.fr/prodom.html`
Email	`prodom@toulouse.inra.fr`
Anonymous FTP	`ftp.toulouse.inra.fr`
Description	The ProDom protein domain database is a useful aid to the analysis of protein domain structure (Sonnhammer and Kahn, 1994). It is a compilation of homologous domains detected in the SWISS-PROT database. Entries in ProDom are individual domains excised from surrounding sequences and stored both as multiple alignments and consensus sequences.

PROSITE

WWW	`http://expasy.hcuge.ch/sprot/prosite.html`
Gopher	`gopher.nih.gov`; go to the "Molecular Biology Databases/ Other Databases/" directory.
Anonymous FTP	`expasy.hcuge.ch`
Description	PROSITE (Bairoch, 1992) is a dictionary and database of protein sites and patterns related to SWISS-PROT. With appropriate computational tools PROSITE can be used to determine if a protein belongs to one of the known families. The WWW site offers search facilities, or a copy of the entire database can be obtained by FTP for local use with several software packages.

REBASE (Restriction Enzyme Data Base)

WWW	`http://www.neb.com/rebase/rebase.html`
Gopher	`gopher.gdb.org`; go to the "Genbank, PIR, Swiss_PROT and other Database Searches/" directory.
Anonymous FTP	`vent.neb.com`
Description	The Restriction Enzyme Database is a collection of information

about restriction enzymes, methylases, the microorganisms from which they have been isolated, recognition sequences, cleavage sites, methylation specificity, the commercial availability of the enzymes, and references (Roberts and Macelis, 1993). It is updated daily, and monthly data releases are available either from the FTP site or by Email. Datafiles compatible with a variety of software packages can be retrieved via anonymous FTP, or by registering for a monthly Email service (contact `macelis@neb.com`).

SBASE

WWW   `http://www.icgeb.trieste.it`
Gopher   `icgeb.trieste.it`
Email   `sbase@icgeb.trieste.it`
Anonymous FTP `icgeb.trieste.it`
Description   SBASE (Pongor *et al.*, 1994) is a collection of annotated protein domain sequences that can be used for homology searching by FASTA and BLAST. Each SBASE entry represents a structural, functional, ligand-binding or topogenic protein segment as defined by the submitting author. The Email server carries out automated searching of SBASE by BLAST.

TFD (Transcription Factor Database)

WWW   `http://dapsas.weizmann.ac.il/bcd/bcd_parent/databanks/tfd.html`
Gopher   `gopher.nih.gov`; go to the "Molecular Biology Databases/Other Databases/" directory
Email   `tfdhelp@ncbi.nlm.nih.gov`
Anonymous FTP `ncbi.nlm.nih.gov`
Description   A relational database of transcription factor recognition elements (Ghosh, 1990) that contains a wealth of information about the proteins, the sites they recognize, the methods by which these sites were established, literature references, related clones and so on. The TFD can be searched by keywords using gopher, but for effective use of the database you must retrieve a version and mount it on your own computer, along with appropriate software. This combination allows you to detect putative binding sites for transcription factors in any given DNA sequence, and to obtain binding site data and published references for each factor.

One example of a program that uses TFD data is SIGNALSCAN (Prestidge, 1991). It can be obtained by sending an Email message to `bioserve@t10.lanl.gov` containing only the phrase `signal-scan` in the text of the Email message. The recent versions of the program incorporate several data sets, including TFD and TRANSFAC (see below). In addition to SIGNALSCAN (which is available for several platforms), a number of other packages can be used with TFD.

TRANSFAC (Transcription Factors and Sites)

WWW `http://transfac.gbf-braunschweig.de/welcome.html`

Email `transfac@gbf-braunschweig.de`

Description TRANSFAC is a database of eukaryotic cis-acting regulatory DNA elements and trans-acting factors (Wingender, 1994). The WWW site offers searching of sites and factors arranged in various categories. Searching can also be performed by SRS at several other sites including: EMBL; `http://www.embl-heidelberg.de/srs/srsc`, and EBI; `http://www.ebi.ac.uk/srs/srsc/`. Help requests and general enquiries should be directed to the Email address.

Pythia (Human Repeat Sequence Analysis)

Email `pythia@anl.gov`

Description An electronic mail server that permits detection and analysis of repeat sequences in submitted human DNA sequences (Milosavljevic and Jurka, 1993). Send a message containing the word "help" in the subject line to the above Email address.

Human and Mouse Genome Project Resources

There are many Internet resources relating to the human genome project. Here are some examples of searchable databases of human and mouse genome mapping resources, including YAC's, STS's, various maps and other resources of use to genome researchers.

Baylor College of Medicine Human Genome Centre

WWW `http://gc.bcm.tmc.edu:8088/`

Description The home page of the Baylor College of Medicine Human Genome Center. Allows searching of Baylor YAC and contig data, and human genome data from other sites. An excellent guide to genetics and human genome resources is located at: `http://gc.bcm.tmc.edu:8088/bio/hgp-resource-guide.html`.

CEPH-Généthon

WWW `http://www.cephb.fr/bio/ceph-genethon-map.html`

Email `ceph-genethon-map@cephb.fr`

Anonymous FTP `ceph-genethon-map.cephb.fr`

Description   Databases relating to the CEPH -Généthon integrated human gene
                    map, allowing (amongst other things) on-line searches of the YAC
                    and STS data used to build a first generation physical map of the
                    human genome (Cohen *et al.*, 1993).

Notes: Information about searching by Email can be found at: `http://`
`www.cephb.fr/bio/mail.html`. See entry for Généthon (below).

CHLC (Cooperative Human Linkage Centre)

WWW   `http://www.chlc.org/`
Gopher   `gopher gopher.chlc.org`
Email   `info-server@chlc.org`
Anonymous FTP   `ftp ftp.chlc.org`
Description   The Cooperative Human Linkage Center provides human genome
                    data and links to several other resources. The data includes
                    information on maps of CHLC markers, and integrated maps
                    combining data from several other mapping groups including
                    Généthon (see entry below).

ELSI (Ethical, Legal and Social Implications)

WWW   `http://www.ncgr.org/elsi/scopenotes.hp.html`
Description   Information relating to the Ethical, Legal and Social Implications
                    (ELSI) program, that deals with the effects of biotechnological
                    progress, particularly the Human Genome Project, on society and
                    the individual.

EUCIB (The European Collaborative Interspecific Mouse Backcross)

WWW   `http://www.hgmp.mrc.ac.uk/MBx/MBxHomepage.html`
Email   `EUCIB-Support@hgmp.mrc.ac.uk`
Description   EUCIB provides an international facility for the high resolution
                    genetic mapping of the mouse genome. One of the key project
                    resources is a 1,000 animal interspecific backcross between C57BL/
                    6 and *Mus spretus*, for which DNA samples, either on filters or as
                    small aliquots, are available.

Généthon

WWW   `http://www.genethon.fr/genethon_en.html`
Gopher   `gopher.genethon.fr`
Email   `info-gopher@genethon.fr`
Anonymous FTP   `ftp.genethon.fr`
Description   Various sources of information relating to the human genome

(Dausset *et al.*, 1990), including data for the 1993-94 Généthon human genetic linkage map (Gyapay *et al.*, 1994). Home of the experimental WWW server GenomeView which integrates access to a variety of human physical and genetic map datasets: `http://ceph-genethon-map.genethon.fr/GenomeView.html`. You can query the WWW server for information on a clone or an STS, or you can use the GenomeView's Email server (send a message with "help" as the subject to `genomeview@genethon.fr` for a complete description).

GenomeNet (Japan)

WWW	`http://www.genome.ad.jp/`
Gopher	`gopher.genome.ad.jp`
Anonymous FTP	`ftp.genome.ad.jp`
Email	`www@genome.ad.jp`
Description	Home of DBGET (see above) and a site with good information about the genome project in general, and about Japanese efforts in particular. Links to many other genome and molecular biology sites.

HGMP (UK MRC Human Genome Mapping Project Resource Centre)

WWW	`http://www.hgmp.mrc.ac.uk/`
Gopher	`gopher.hgmp.mrc.ac.uk`
Email	`biohelp@hgmp.mrc.ac.uk`
Description	A key Human Genome Project site in the UK, that develops and makes available databases of genes, genetic markers and map locations. The Centre also provides biological materials (clones, probes, primers, libraries, cell lines, FISH mapping) relating to the human and mouse genome projects. For example, the primer bank contains many primers for human disease loci, and for human and mouse mapping, and the probe bank contains information on relevant probes. Links to EHCB,CAD and various other databases are also available through this site

Introductory Human Genome Material

WWW	`http://outcast.gene.com/ae/AB/UT/HGP/index.html` `http://www.gdb.org/Dan/DOE/intro.html`
Gopher	`merlot.gdb.org`; go to the "Genome Project/" directory.
Description	These sites offer good, non-technical overviews of the human genome project and related issues. Useful for teaching and lay education purposes.

LLNL Human Genome Center

WWW   http://www-bio.llnl.gov/bbrp/genome/
                       genome.html

Description   The Lawrence Livermore National Laboratory Human Genome
                       Center develops biological, bioinformatic, and technological
                       resources for human genome mapping and sequencing, details of
                       which are available at this site.

MIT/Whitehead Centre for Genome Research

WWW   http://www-genome.wi.mit.edu/

Description   WWW server for the Center for Genome Research at the Whitehead
                       Institute for Biomedical Research in Cambridge, Massachusetts,
                       USA. It contains information on human and mouse genome maps,
                       and provides genome analysis software.

Oak Ridge National Laboratory

WWW   http://www.ornl.gov/TechResources/
                       Human_Genome/home.html

Description   A service sponsored by the U.S. Department of Energy Human
                       Genome Program, with general information and numerous relevant
                       Internet links. Includes access to and searching of on-line editions
                       of Human Genome News, the publication of the U.S. Human
                       Genome Program, and access to documents from The Human
                       Genome Management Information System.

Portable Dictionary of the Mouse Genome

WWW   http://mickey.utmem.edu/front.html
Anonymous FTP   ncbi.nlm.nih.gov or nb.utmem.edu

Description   The Portable Dictionary of the Mouse Genome is a useful
                       compendium of mouse genome information extracted from various
                       on-line and published mouse genetics resources (Williams,1994).
                       A guide to the dictionary is at: http://mickey.utmem.edu/
                       description/Guide.html. It is designed for use on
                       computers without network access, it fits on a small number of
                       diskettes, and the data can be used in several different database
                       programs in common use. The files are intended to be downloaded
                       either from the WWW site or by FTP, for use on local computers.

SIGMA (System for Integrated Genome Map Assembly)

WWW   http://www.ncgr.org/sigma/intro.html
Description   A graphical database tool for building and viewing integrated

genome maps. Sigma allows the user to integrate data from a variety of sources into a single map, generate reports, and perform complex analyses of the data. Requires a forms-capable WWW browser.

University of Utah

WWW   `http://www.genetics.utah.edu//`
Gopher   `corona.med.utah.edu`
Description This site offers, amongst other things, a searchable listing of a large collection of sequence tagged sites (STS), arranged by human chromosome. Each entry contains primer sequences, details on allele sizes, and PCR and electrophoresis conditions.

Washington University EST (Merck) Database

WWW   `http://genome.wustl.edu/est/est_general/ est_project_intro.html`
Description The Washington University/Merck EST (Expressed Sequence Tag) project home page. Sequences of the ESTs are lodged at NCBI in dbEST (see above).

Biological Reagent Sources

This section outlines a number of sites, not already covered in the above sections, that provide access to reagents that may be useful in your work. Some of these organizations also publish hard-copy catalogues, but the advantage of the Internet sites is that they can be searched by keywords, they are more up to date than the catalogue, there are minimal expenses involved, and they don't take up space on your bookshelf! More and more companies are developing Internet capability; links to many of these can be found at the HUM-MOLGEN home page (see below) and at: `http:// 155.41.115.114/jumper/biolcomm.html` or `http://www.atcg.com/ aguide//comppage/comppage.htm`.

ATCC (The American Type Culture Collection)

WWW   `http://www.atcc.org/`
Gopher   `culture.atcc.org`
Email   `help@atcc.org`   (for comments or help)
             `request@atcc.org`
Telnet   `culture.atcc.org`
Description The ATCC database and on-line searchable catalogues. Contains information on cell lines, microbes, recombinant DNA libraries and clones, and oligonucleotides related to genome research.

Notes: For Telnet login as `search` and use the password `common`.

ATCG (Anderson's Timesaving Comparative Guides on the Web)

WWW   http://www.atcg.com/aguide//atcghome.htm
Email   mailto@ATCG.com
Description   This well-organized site provides easy access to product information in biology and chemistry. The stated aim of this organisation is "to enable researchers to spend more time at the bench because they spend less time with catalogs." The information covered at present includes commercially available restriction and modifying enzymes, recombinant libraries, filters and membranes, and links to many suppliers with Internet connections.

Cedars-Sinai Molecular Genetics

WWW   http://www.csmc.edu/genetics/korenberg/
               korenberg.html#B
Description   Provides an integrated YAC/BAC/PAC resource for genome mapping, from which reagents such as mapped bacterial artificial chromosome (BAC) and P1 artificial chromosome (PAC) clones can be obtained. Source of mapping information, with emphasis on human chromosome 21 and its associated phenotypic abnormalities.

Coriell Cell Repository

Telnet   Coriell.umdnj.edu. (login "online").
Description   The Coriell Cell Repository distributes resources from The National Institute of General Medical Sciences, including NIGMS Mapping panels, DNA and cell cultures for human and rodent somatic cell hybrids, a large variety of human and animal cells (and DNA), and a wealth of information on the cells including references. On-line, searchable catalogues for the repository are available by Telnet.

ECACC (European Collection of Animal Cell Cultures)

Gopher   gopher.gdb.org "Search Databases at Hopkins..." directory
Description   The European Collection of Animal Cell Cultures (Porton Down, U.K.) searchable databases of cell lines, hybridomas and DNA probes, including the contents of the European Human Cell Bank. Fully described at: gopher://merlot.gdb.org/0R1123-3519-/Database-local/cultures/ecacc/about/info/info-1.data

E. coli Genetic Stock Center at Yale University

WWW   gopher://cgsc.biology.yale.edu/11/CGSC
Gopher   cgsc.biology.yale.edu

Email   `mary@cgsc.biology.yale.edu`
Description   A database of *E.coli* genome and strain information, including *E.coli* genetic map data, and details of many strains, mutations, genes,and references. Useful for locating strains with specific mutations or genotypic combinations, as well as for examining stock center information on mutations and genes. Contact Email address for details of telnet access.

HyperCLDB (Cell Line Database)

WWW   `http://www.ist.unige.it/cldb/indexes.html`
Gopher   `gopher.ist.unige.it`
Telnet   `istge.ist.unige.it`
Description   HyperCLDB is a database containing information on about 3000 human and animal cell lines that are available from many Italian laboratories and from various European cell banks and cell culture collections. This has resulted from the Interlab Project, a description of which is at: `http://www.ist.unige.it/interlab/intro.html`.

IMAGE Consortium

WWW   `http://www-bio.llnl.gov/bbrp/image/image.html`
Email   `info@image.llnl.gov`
Description   The IMAGE (Integrated Molecular Analysis of Genomes and their Expression) Consortium of several international laboratories arranges for the sharing and distribution of high quality, arrayed human cDNA libraries for which sequence, map and expression data are publically available. The aim is to develop a "master array" which contains a representative cDNA clone from every human gene.

Jackson Laboratory Stocks

WWW   `http://www.jax.org/`
Gopher   `hobbes.informatics.jax.org`
Description   The Jackson Laboratory offers a collection of diverse mouse strains including inbred strains, recombinant strains, congenic strains, and many mutant stocks. Genomic DNA extracted from many of their mice stocks is also available as is a price list of all of these items. Nomenclature guides, other information on mouse genetics, and links to various bioinformatics resources are also provided.
DNA   `http://www.jax.org/resources/documents/dnares`
Pricelist   `http://www.jax.org/resources/documents/pricelist/index.html`

RLDB (Reference Library Database)

WWW	`http://gea.lif.icnet.uk/`
Gopher	`ftp.embl-heidelberg.de`
Email	`genome@icrf.icnet.uk`
Anonymous FTP	`ftp.embl-heidelberg.de`
Description	The Reference Library System (Zehetner and Lehrach, 1994) is a model for sharing recombinant DNA clones and pooling the experimental data derived with these materials, leading to a powerful database for gene mapping, gene cloning and gene discovery purposes. Cosmid, YAC, P1 and cDNA libraries (from which the data in RLDB are derived) are arrayed on filters and made available to the research community. RLDB and RLDB2 (a newer version) stores the hybridization and PCR results for these clones generated by the research community. Searches of RLDB2 return files containing hypertext links to other related information, such as data in OMIM, GDB etc.

General Information Sources

Amino Acid Data

WWW	`gopher://gopher.imb-jena.de/00/ftp/images/PROTEINS/amino_acids/amino_acid.txt`
Description	Useful tables listing several properties of amino acids and their single letter codes. Amino acid images with atomic numbering can be viewed at `gopher://gopher.imb-jena.de/11/ftp/images/PROTEINS/amino_acids`.

EBI (European Bioinformatics Institute)

WWW	`http://www.ebi.ac.uk/`
Gopher	`gopher.ebi.ac.uk`
Anonymous FTP	`ftp.ebi.ac.uk`
Description	The servers of the European Bioinformatics Institute (EBI), located near Cambridge in the UK (Emmert *et al.*, 1994). The EBI is an Outstation of EMBL that provides bioinformatics databases, software and other resources, and links to relevant sites. EBI has taken over all the bioinformatics activities previously located at the EMBL Data Library in Heidelberg.

EMBL File Server

WWW site	`http://beta.embnet.unibas.ch/embnet/html/embl.html`

Email   `NetServ@EMBL-Heidelberg.DE`
Description   The EMBL File Server allows external users to request files by electronic mail. Many data and software files, including REBASE in various formats, TFD, and the EMBL nucleotide sequence and protein databases are available. For more information, send a mail message to the above address containing only the word "HELP" in the body of the message.

FTP Archives for Molecular Biology

WWW   `gopher://genome-gopher.stanford.edu/11/ftp`
Gopher   `genome-gopher.stanford.edu`
Description   A comprehensive listing of FTP sites for databases, software, documents and other resources.

Genetic Linkage Analysis Home Page

WWW   `http://lenti.med.umn.edu/linkage/linkage.html`
Gopher   `lenti.med.umn.edu;` " Biologically Related Information/ Genetic Linkage Analysis" directory.
Description   A good place to start looking for information on genetic linkage analysis. Links to relevant databases, software, and literature.

HUM-MOLGEN Home Page

WWW   `http://www.informatik.uni-rostock.de/ HUM-MOLGEN/index.html`
Description   An internet communication forum dealing with many issues relating to human molecular genetics. The homepage is organized into sections such as requests for collaboration, biotechnology and molecular biology, education, diagnostics, computational genetics, ethical and social issues, literature and so on. There are hypertext links to many other resources, and an associated mailing list (HUM-MOLGEN, see above).

IUBIO (Indiana University)

WWW   `http://iubio.bio.indiana.edu/`
Gopher   `iubio.bio.indiana.edu`
Anonymous FTP   `iubio.bio.indiana.edu`
Description   IUBio Archive is an archive of biology data and software that allows browsing of various resources, searching of sequence databases and Bionet News archives, and retrieval of software and data. Has links to other resources and sites.

Johns Hopkins Bio-Informatics Home Page

WWW   `http://www.gdb.org/hopkins.html`
Description   A well organised site with useful links to servers, databases, publications and other resources.

LiMB (Listing of Molecular Biology Databases)

Gopher   `gopher.nih.gov`; in the "Molecular Biology Databases/Other Databases/" directory.
Email   `limb@life.lanl.gov`
Description   LiMB contains information about the contents and maintenance of molecular biology databases (Keen *et al.*, 1992). It is well worth searching when you want to locate a database, or determine what sort of data it stores and how it can be obtained.

Molecular Biologist's Desk Reference

WWW   `http://molbio.info.nih.gov/molbio/desk.html`
Description   Useful homepage with links to tables for genetic code and codon usage; amino acid properties, structures, and nomenclature; and molecular biology protocols and bibliographic services.

Mouse and Rat Research Home Page

WWW   `http://www.cco.caltech.edu/~mercer/htmls/rodent_page.html`
Description   An excellent place to find almost anything to do with mice (mainly). Many links to relevant sites, dealing with genomics, transgenics, general biology of rodents, animal husbandry, literature, meetings and so on. Comprehensively links rodent resources on the Internet.

Pedro's Biomolecular Research Tools

WWW   `http://www.public.iastate.edu/~pedro/research_tools.html`
Description   An outstanding collection of WWW Links to information and services useful to molecular biologists, including extensive listings of molecular biology database search and analysis facilities, bibliographies, guides, tutorials and journal contents.

The Resource for Molecular Cytogenetics

WWW   `http://rmc-www.lbl.gov/`
Description   The Resource for Molecular Cytogenetics at Lawrence Berkeley Laboratory and the University of California, San Francisco, focuses on the application of molecular cytogenetics in clinical and biological studies. A database of mapped probes for fluorescent *in situ* hybridization analyses (FISH), FISH images and other relevant information are available.

University of Washington Cytogenetics and Genome Resources

WWW   `http://www.pathology.washington.edu/cyto_page.html`
Gopher   `larry.pathology.washington.edu`
Email   `dadler@u.washington.edu`
Description   Source of human and mouse chromosome ideograms, and chromosome scans, in down-loadable formats.

Virtual Library: Biosciences

WWW   `http://golgi.harvard.edu/htbin/biopages`
Description   A searchable index of Internet resources catalogued in different subject headings. Very useful for navigating to particular areas of the Net.

Virtual Library: Genetics

WWW   `http://www.ornl.gov/TechResources/Human_Genome/genetics.html`
Description   A comprehensive listing of genetics resources on the Internet, with links to each site.

Acknowledgements

I am indebted to Gordon Findlay for devoting a disproportionate amount of his time to molecular biology computing needs. This chapter is merely a compilation that samples the work of many dedicated individuals who are the lifeblood of the Internet - without their efforts, none of this would be possible. I am responsible and regretful for any omissions and inaccuracies. My work is supported by the Health Research Council of New Zealand and the New Zealand Lottery Grants Board.

References

Appel R.D., Bairoch A. and Hochstrasser D.F. 1994. A new generation of information retrieval tools for biologists: The example of the ExPASy WWW server. Trends Biochem. Sci. 19: 258-260.

Bairoch, A. 1992. PROSITE: a dictionary of sites and patterns in proteins. Nucleic Acids Res. 20 (Suppl): 2013-2018.

Bairoch, A. and Boeckmann B. 1991. The SWISS-PROT protein sequence data bank. Nucleic Acids Res. 19 (Suppl): 2247-2249.

Benson, D.A., Boguski, M., Lipman, D.J. and Ostell, J. 1994. GenBank. Nucleic Acids Res. 22: 3441-3444.

Bleasby, A., Griffiths, P., Hines D., Marshall, S., Staniford, L., Hoover, K. and Kristofferson, D. 1993. The BIOSCI newsgroups--computer networks changing biology. Trends Biochem. Sci. 18: 310-311.

Bleasby, A.J. and Wootton, J.C. 1990. Construction of validated, non-redundant composite protein sequence databases. Protein Eng. 3: 153-9.

Boguski, M., Lowe, T.M. and Tolstoshev, C.M. 1993. dBEST: Database for "Expressed Sequence Tags". Nature Genet. 4: 332-333.

Bucher, P. and Trifonov, E.N. 1986. Compilation and analysis of eukaryotic POL II promoter sequences. Nucleic Acids Res. 14: 10009-10026.

Cohen, D., Chumakov, I., Weissenbach, J. 1993. A first-generation physical map of the human genome. Nature. 336: 698-701.

Dausset, J., Cann, H., Cohen D., Lathrop, M., Lalouel, J-M. and White, R. 1990. Program description. Centre d'Etude du Polymorphisme Humain (CEPH): Collaborative genetic mapping of the human genome. Genomics. 6: 575-577.

Emmert, D.B., Stoehr, P.J., Stoesser, G., and Cameron, G.N. 1994. The European Bioinformatics Institute (EBI). Nucleic Acids Res. 22: 3445-3449.

Etzold, T. and Argos, P. 1993. SRS an indexing and retrieval tool for flat file data libraries. Comput. Appl. Biosci. 9: 49-57

Fasman, K.H., Cuticchia, A.J. and Kingsbury, D.T. 1994. The GDB(TM) Human Genome Data Base Anno 1994. Nucleic Acids Res. 22:3462-3469.

Ghosh, D. 1990. A Relational Database of Transcription Factors. Nucleic Acids Res. 18: 1749-1756.

Gilbert, W. 1991. Towards a paradigm shift in biology. Nature. 349: 99.

Gyapay, G., Morisette, J., Vignal, A., Dib, C., Fizames, C., Millasseau, P., Marc, S., Bernardi, G., Lathrop, M. and Weissenbach, J. 1994. The 1993-1994 Genethon human genetic linkage map. Nature Genet. 7: 246-339.

Higgins, D.G., Fuchs, R., Stoehr, P.J. and Cameron, G.C. 1992. The EMBL data library. Nucleic Acids Res. 20 (Suppl): 2071-2074.

Keen, G., Redgrave, G., Lawton, J., Cinkosky, M., Mishra, S., Fickett, J. and Burks, C. 1992. Access to molecular biology databases. Mathl. Comput. Modelling. 16: 93-101.

McKusick, V.A. 1994. Mendelian Inheritance in Man : A Catalog of Human Genes and Genetic Disorders. 11th Edition. Johns Hopkins University Press, Baltimore. pp 3009.

Milosavljevic, A., and Jurka, J. 1993. Discovering simple DNA sequences by the algorithmic significance method. Comput. Appl. Biosci. 9: 407-11.

Nadeau, J.H., Grant, P.L., Mankala, S., Reiner, A.H., Richardson, J., and Eppig, J.T. 1995. A rosetta stone of mammalian genetics. Nature. 373: 363-365.

Pearson, P., Francomano, C., Foster, P., Bocchini, C., Li, P. and McKusick, V. 1994. The status of Online Mendelian Inheritance in Man (OMIM) medio. Nucleic Acids Res. 22: 3470-3473.

Pongor, S., Hatsagi, Z., Degtyarenko, K., Fabian, P., Skerl, V., Hegyi, H., Murvai, J. and Bevilacqua, V. 1994. The SBASE protein domain library, release 3.0: A collection of annotated protein sequence segments. Nucleic Acids Res. 22: 3610-5.

Prestridge, D.S. 1991. SIGNAL SCAN: A computer program that scans DNA sequences for eukaryotic transcriptional elements. CABIOS 7: 203-206.

Ritter, O. 1995. IGD: Comprehensive integration of human genome data and analytical tools. HUGO Genome Digest. 2: 10-12.

Roberts, R.J., Macelis, D. 1993. REBASE--restriction enzymes and methylases. Nucleic Acids Res. 21: 3125-3137.

Sidman KE., George, D. G., Barker, W. C. and Hunt, L. T. 1988. The protein identification resource (PIR). Nucleic Acids Res . 16: 1869-71.

Sonnhammer, E.L. and Kahn, D. 1994. Modular arrangement of proteins as inferred from analysis of homology. Protein Sci. 3: 482-92.

Williams, R.W. 1994. The portable dictionary of the mouse genome: A personal dataabse for gene mapping and molecular biology. Mammalian Genome. 5: 372-375

Wingender, E. 1994. Recognition of regulatory regions in genomic sequences. J. Biotechnol. 35:273-280.

Woychik, R.P., Wassom, J.S., Kingsbury, D. and Jacobson, D.A. 1993. TBASE: A computerized database for transgenic animals and targeted mutation. Nature 363: 375-6.

Further Reading

Zehetner, G., Lehrach, H. 1994. The Reference Library System : Sharing biological material and experimental data. Nature. 367: 489-491.

From: *Internet for the Molecular Biologist.*
ISBN 1-898486-02-6 ©1996 Horizon Scientific Press, Wymondham, U.K.

12

INTERNET RESOURCES FOR FUNGI

Kathie T. Hodge

Introduction

Resources of particular interest to those studying fungi are listed below. The two most comprehensive starting points are the branches of the WWW Virtual Library covering mycology and yeasts.

Internet Resources

American Phytopathological Society

WWW   http://www.scisoc.org/

Description   The web site of the American Phytopathological Society (APSnet) includes many resources of interest to Plant Pathologists, including information on APS membership, annual meetings, the APS annual report, career and graduate school information, the APS Press on-line catalogue, book reviews, abstracts of articles in APS journals, titles of newly accepted journal articles, and a valuable compilation of common names of plant diseases.

Notes: Subscription ($25 per year for APS members) provides access to additional services, including updated job listings, searchable annual meeting program, abstracts from Division meetings, a directory of APS members, and access to discussion groups and electronic symposia.

American Type Culture Collection

WWW   http://www.atcc.org/

Description   Catalogues of the American Type Culture Collection (ATCC) can be searched for fungi and other organisms. On-line ordering is available.

BIONET Groups

News   `bionet.mycology; bionet.molbio.yeast`
Description The BIOSCI/bionet discussion groups bionet.mycology and bionet.molbio.yeast are of particular interest to those interested in fungi. See Chapter 9 for more information.

British Society for Plant Pathology

WWW   `http://www.scri.sari.ac.uk/bspp/`
Description The BSPP Web page includes information on the Society, the BSPP Newsletter, and a guide to the terminology of plant pathology.

Candida albicans **information**

WWW   `http://alces.med.umn.edu/Candida.html`
Description The Web page of the *Candida albicans* mapping project at the University of Minnesota. Among the available information are physical maps of nuclear and mitochondrial genomes, gene sequences (including unpublished information), results of gene disruptions, PCR primer sequences, probes, catalogues of strains, programs for sequence manipulation and searching that can be run remotely, a directory of researchers' Emailaddresses, and the Candida News electronic newsletter.

CBS Culture Collection

WWW   `http://www.cbs.knaw.nl`
Description On-line catalogue of the culture collection of the Centraalbureau voor Schimmelcultures (Baarn, The Netherlands). Bacteria, yeasts, and filamentous fungi can be searched by name or by their special properties.

Fungal Genetics Stock Center

WWW   `http://kumc.edu/research/fgsc/main.html`
Description The Fungal Genetics Stock Center at the University of Kansas Medical Center provides catalogues of strains of *Aspergillus*, *Fusarium*, *Neurospora*, and *Sordaria*, as well as cloned genes and genetic libraries from these fungi. Also available here is the Fungal Genetics Newsletter, including full text and figures.

Fungal Genome Resource

WWW   `http://fungus.genetics.uga.edu:5080/`
Description The Fungal Genome Resource at the University of Georgia includes physical maps of *Aspergillus nidulans* and information on ODS (Ordering DNA Sequences) physical mapping software for Solaris and Windows systems.

Inoculum: The MSA Newsletter

Gopher   `gopher://nmnhgoph.si.edu:70/11/.botany/.myco/
             .inoculum`
Description The archives of Inoculum, the newsletter of the Mycological Society of America, are available for browsing and searching.

Microbial Germplasm Database

Gopher   `gopher://gopher.bcc.orst.edu/11/mgd`
Description This gopher server provides catalogues for many culture collections, including fungi and other organisms.

Microbial Strain Data Network

WWW   `http://www.bdt.org.br/msdn/msdn.html`
Description A large number of culture collections of microorganisms can be searched through the Microbial Strain Data Network Web page.

Notes: For help or information send Email to `MSDN@phx.cam.ac.uk`

MSA Bulletin Board

Gopher   `gopher://huh.harvard.edu/1m/project_information/
             msa-bbs`
Description The electronic bulletin board of the Mycological Society of America includes job listings and announcements of interest to mycologists.

Notes: To post a message, or for help, send Email to `MSA-news@huh.harvard.edu`

Mycologists Online

Gopher   `gopher://muse.bio.cornell.edu:70/11/directories/
             mo`

Description   Mycologists Online lists Email and snail mail addresses of mycologists,
              lichenologists, and collection curators throughout the world.

Notes: For information or inclusion contact Dr. Pavel Lizon at `PL18@cornell.edu`

Saccharomyces Genomic Information Resource

WWW   `http://genome-www.stanford.edu/`
Description   Complete chromosome sequences, genetic and physical maps, and access
              to yeast protein and gene databases are available at the *Saccharomyces*
              Genome Database at Stanford University.

U.S. National Fungus Collections Databases

Telnet   `fungi.ars-grin.gov` (enter "login user" then use the password
              "user")
Description   This very important site run by the US Dept. of Agriculture, Agricultural
              Research Service serves searchable data from Fungi on Plant and Plant
              Products in the US, including host-fungus distributions, literature, and
              plant common names. Also available are databases of references for
              systematic mycology and plant pathology, the Mycological Society of
              America membership directory, specimens in the National Fungus
              Collections Herbarium, a searchable index to Saccardo's Sylloge
              Fungorum, and the Index of Fungi, volumes 1 to 4.

Notes: Also available at this site are databases of the Animal and Plant Health Inspection
Services (APHIS): type "login APHIS" and use the password APHIS. For help send a
message to `dave@fungi.ars-grin.gov`

World Data Center on Microorganisms

WWW   `http://www.wdcm.riken.go.jp/`
Description   The World Federation of Culture Collections web page includes
              searchable indices of over 400 registered culture collections in more
              than 50 countries worldwide, as well as directories of Email addresses
              of collection staff.

World-Wide Web Virtual Library: Mycology

WWW   `http://muse.bio.cornell.edu/taxonomy/fungi.html`
Description   All known internet resources of interest to professional mycologists are
              summarized at this regularly updated site.

World-Wide Web Virtual Library: Yeast

WWW   `http://genome-www.stanford.edu/VL-yeast.html`

Description   A summary of internet resources of particular interest to researchers of yeast (*Saccharomyces cerevisiae*, *Schizosaccharomyces pombe*, and *Candida albicans*).

Yeast Researchers: On-Line Directory

Gopher   `gopher://merlot.gdb.org/11/biol-search/yeast-Email`

Description   A directory of Email addresses of "yeast interested people."

From: *Internet for the Molecular Biologist.*
ISBN 1-898486-02-6 ©1996 Horizon Scientific Press, Wymondham, U.K.

13

INTERNET RESOURCES FOR INVERTEBRATES

Steven J.M. Jones and David Hodgson

C. elegans

Avery Lab

WWW   http://eatworms.swmed.edu/Worm_labs/Avery/index.html

Description Holds comprehensive information on how to set up a miniature electropharyngeogram (EPG) apparatus as well as a laser ablation system. There is also some detailed information on the physiology of the *C. elegans* pharynx with a clickable atlas of the pharynx. This site also has software which allows the viewing of electrophysiology data.

Vancouver Worm Community

WWW   http://genekit.medgen.ubc.ca/index.html

Description The front end to several *C. elegans* labs in British Columbia. Provides information on some of the genetic balancers which are available for *C. elegans*. But more specifically, this site provides information of the Cosmid Transgenics from the Canadian Genome Analysis and Technology Project (CGAT). This database consists mainly of cosmid transgenic strains made from cosmids used in the *C. elegans* genome sequencing project and information on how these strains can be obtained is also given.

Blumenthal Lab: The Worm Web

WWW   http://hyrax.bio.indiana.edu:8080/

Description This site contains the "Universal Worm Term Looker-Upper". Which allows the searching of various terms from a number of nematode specific and non-specific databases. Also has a useful collection of protocols and methods used in *C. elegans* biology as well as a comprehensive collection of abstracts and Worm Breeders Gazette articles.

Bowerman Lab

WWW   `http://eatworms.swmed.edu/Worm_labs/Bowerman/`
Description   Collection of browser launchable quicktime movies of rotational series
             of the pharynx using confocal microscopy.

The Community Systems Laboratory: The Worm Community System

WWW   `http://csl.ncsa.uiuc.edu/`
Description   The worm community system (WCS) is a digital library for *C. elegans*.
             This site provides an online user guide for this software. This also
             provides the information required to get and install WCS on your
             workstation.

The Netherlands Worm Home Page

WWW   `http://eatworms.swmed.edu/Worm_labs/Plasterk/`
             `homepage/shakes.html`
Description   This is the site to find out about reverse genetics in *C. elegans*. A
             comprehensive document on this subject is provided as well as details
             of the protocols which need to be followed.

Notes: There is also a database of transposon (Tc1) end sequences or tags created by
this laboratory which is searchable on-line or can be down-loaded.

Dauer World

WWW   `http://dauerdigs.biosci.missouri.edu/Dauer-`
             `World/index.html`
Description   A *C. elegans* site specialising in the genetics of dauer formation and
             again in *C. elegans*.

The *C. elegans* Genome Project World Wide Web Site - Sanger Centre

WWW   `http://www.sanger.ac.uk/`
Description   The European site for the *C. elegans* genome sequencing project.
             Containing sequences generated by the sequencing consortium. An on-
             line blast server is available to allow easy searching of the C. elegans
             nucleotide sequences. Wormpep, the *C. elegans* protein database is
             also available as well as a small collection of confirmed splice sites. A
             tutorial on how to use the genefinder program within the *C. elegans*
             database ACeDB is also provided.

Washington Universtiy: Genome Sequencing Center, Washington University

WWW   `http://genome.wustl.edu/`
Description The North American site for the *C. elegans* genome sequencing project. Like the European site sequences derived from the sequencing project can be browsed and downloaded. Also available is a *C. elegans* blast database consisting of the top 25 non-overlapping blastx hits to the PIR protein database for each of the sequenced *C. elegans* cosmids. This site also contains a study of repetitive sequences in *C. elegans* and provides interactive forms which can be used to search for repeat families.

Caenorhabditis Genetics Centre

Gopher   `gopher://elegans.cbs.umn.edu/1`
Description Not fully in html form as yet, but provides an essential resource for *C. elegans* biologists being the repository of more than 2000 genetics strains, including one allele of each mapped gene. This site includes a strain list for *C. elegans* as well as an extensive *C. elegans* bibliography, both of which can be searched online using simple keywords. This site also contains the bionet.celegans newsgroup archive.

A *Caenorhabditis elegans* DataBase

WWW   `http://moulon.inra.fr/acedb/acedb.html`
Description This provides an online version of the database program ACeDB created by Richard Durbin and Jean Thierry-Mieg. Uses a combination of the ACeDB query language and the graphical output of ACeDB to allow efficient use of the internet. Instructions for searching and a brief description of the query syntax are provided. Invaluable for those without their own implementation of ACeDB.

The Hope Lab

WWW   `http://eatworms.swmed.edu/Worm_labs/Hope/`
Description Contains an expression pattern database for many *C. elegans* genes. This image collection of expression patterns currently contains only expression patterns derived from beta-galactosidase-reporter fusions. It is promised that the database will be complemented with expression patterns derived from both *in situ* hybridisation and also immunohistochemical staining methodologies.

Drosophila

FlyBase

WWW   `http://morgan.harvard.edu/`
Gopher   `gopher://ftp.bio.indiana.edu/hh/Flybase/`
FTP   `ftp://flybase.bio.indiana.edu`
Description   Flybase is the main *Drosophila* internet resource for molecular and genetic data. It holds very large amounts of information on genes, mutations, maps, stocks, researchers, citations, database entries and search tools to access this data. Under development are graphical maps for browsing data by cytogenetic/genetic location. It makes good use of the WWW interface and is expanding all the time.

Note: The archive for `bionet.drosophila` is also held here.

FlyView

WWW   `http://pbio07.uni-muenster.de/`
Description   Flyview is a *Drosophila* development and genetics image database. It specialises in gene expression patterns, such as enhancer trap lines and cloned genes. Although it is still under-construction, it currently serves pictures from various developmental stages via its database search and retrieval page. The images are accompanied by text descriptions which can in turn be used as search keywords.

Flybrain

WWW, with sites in;
    USA   `http://flybrain.neurobio.arizona.edu/`
    UK   `http://flybrain.gla.ac.uk/`
    Germany `http://flybrain.uni-freiburg.de/`
Description   Flybrain describes itself as "An Online Atlas and Database of the *Drosophila* Nervous System". Still in its infancy, Flybrain makes good use of the hypertext environment to allow you browse though various aspects and information all related to neuroanatomical data and forms a basic atlas to the adult central brain.

Gerard Manning's World-Wide Web Virtual Library Drosophila Page

WWW   `http://www-leland.stanford.edu/~ger/`
              `drosophila.html`
Description   Definitive collection of *Drosophila* related WWW links, a good starting point for *Drosophila* site surfing.

The Hartl Lab, Harvard University, Cambridge, MA.

WWW   `http://golgi.harvard.edu/hartl/`
Description   This laboratory is involved with the *Drosophila* genome mapping project. There is information about this and members of the lab as well as a list of their publications.

The Nusslein-Volhard Lab, Tubingen, Germany

WWW   `http://wwweb.mpib-tuebingen.mpg.de/Abt.3/`
Description   Information can be found here on publications from the lab as well as details of their 1,700 large *Drosophila* stock collection.

The Fischbach Lab, Freiburg, Germany.

WWW   `http://deep-thought.biologie.uni-freiburg.de/data/kff.html`
Description   Information on their work on the "Genetic basis of brain development and function" can be found here. Also includes lists of their derived mutants as well as their publications and a permanent on-line poster session.

J.R. Jacobs, McMaster University

WWW   `http://www.science.mcmaster.ca/Biology/faculty/Jacobs.html`
Description   A concise summary of the work being carried out at this lab on genetic and morphological dissection of cell determination and cell to cell recognition in the *Drosophila* embryo and within tissue culture.

Kim Kaiser's Lab at Glasgow University

WWW   `http://macserver.molbio.gla.ac.uk/IBLS/staff/k-kaiser.html`
Description   Entitled "Molecular Genetic Analysis of Brain Structure and Function in *Drosophila*", these pages provide information about staff at the lab and their individual research projects.

Penn State University.

WWW   `http://millreef.mcb.psu.edu/deptpage/faculty.htm`

Description   Many of the members of the Faculty of Biochemistry and Molecular
               Biology are involved in *Drosophila* research, links to their individual
               pages are provided here. Many research fields are represented including,
               population genetics, muscle development and evolution.

The Nagoshi Drosophila Lab

WWW   `http://fly2.biology.uiowa.edu/`
Description   Details here of their work on oogenesis, enhancer trap lines and other
               related research. There is also a page of computer generated, ray traced
               pictures and movies to download.

The *Drosophila* Newsgroup

news   `bionet.drosophila`
Description   The Usenet discusson forum for *Drosophila* related postings and issues.

Other Invertebrates

Mosquito Genomics WWW Server

WWW   `http://klab.agsci.colostate.edu/`
Description   This represents the canonical site for all work on mosquito molecular
               biology. Provided is an on-line searchable and browsable version of the
               moquito genome database (AaeDB). Particularly useful are graphical
               images of various genomic maps, included at present is a genetic map,
               an RFLP linkage map and also a FISH physical map.

Entomology on World-Wide Web (WWW)

WWW   `http://www.colostate.edu/Depts/Entomology/`
               `www_sites.html`
Description   Contains a comprehensive list of entomology sites. Although the vast
               majority of the topics and sites listed are not molecular biology related it
               does provide a useful jumping-off point for those interested in the world
               of insect research. Sites listed include ones specialising in beekeeping,
               forest pests, lepidoptera, arachnology and even a link to "Buzz the fly",
               an entomological comic strip.

From: *Internet for the Molecular Biologist.*
ISBN 1-898486-02-6 ©1996 Horizon Scientific Press, Wymondham, U.K.

14

INTERNET RESOURCES FOR PLANT MOLECULAR BIOLOGY

S. Beckstrom-Sternberg, G. Juvik, D. Bigwood, J. Barnett,
J. Krainak, M. Sikes, J. Martin, M. Shives, S. Cartinhour, S. Heller
and J.P. Miksche

Part I. Genome Databases

Agricultural Genome Information Server (AGIS)

WWW	`http://probe.nalusda.gov/`
Gopher	`probe.nalusda.gov`
Anonymous FTP	`probe.nalusda.gov`
Description	The Agricultural Genome Information Server (AGIS) server provides genome databases of agriculturally important plant and animal (Sept. 1995) species, and model organisms. The plant species include *Arabidopsis*, common bean and soybean, *Chlamydomonas*, cotton, small grains (wheat, barley, rye, etc.), maize, rice, tomato, potato, pepper, sorghum, and forest trees. Also included is Mendel, the database from the Commission on Plant Gene Nomenclature.

The WWW versions of the databases are provided in ACeDB format and are browsable via hypertext links. They are also searchable via WAIS, fuzzy searching (agrep), and querying through ACeDB (query by example, query builder, and ACeDB query language). The databases are characterized by links to external databases, including EMBL, Enzyme Data Bank, Genbank, Genobase, GRIN (Germplasm Resources Information Network), Kyoto Ligand Database, Medline, PDB (Protein Database), PIR (Protein Information Resource), Prosite, Rebase, Selkov Enzyme and Metabolic Pathways, and SwissProt, and tools are provided to enhance and augment external connections. Documentation on the server includes a full suite of ACeDB documentation, electronic newsletters, journals, and other publications including: Plant Genome meeting abstracts, Weeds World, the Rice Genetics Newsletter, Report of the Tomato Genetic Cooperative, and the announcement for the new peer-reviewed electronic journal, The Journal of Quantitative Trait Loci.

Notes: If you do not have a WWW browser, but have telnet capability, use the lynx, text-based browser. Telnet to: `probe.nalusda.gov` and use the login "lynx" (no

quotes). No password is necessary. You MUST use vt100 terminal emulation in order to use lynx. Also, lynx has been modified so you can only browse the AGIS along with a few related servers. As a novel extension of ACeDB for non-genome databases, several plant databases are provided, including PhytochemDB (plant phytochemicals), EthnobotDB (world wide plant uses), FoodplantDB (Native American food plants), and PVP (Plant Variety Protection - soybean subset). The AGIS is a service provided by the U.S. Department of Agriculture's National Agricultural Library in Beltsville, Maryland, USA. Funding is provided through the USDA, Agricultural Research Service, National Genetic Resources Program. Send comments to:
`feedback@probe.nalusda.gov.`

Send requests for help to   `help@probe.nalusda.gov`

> Genome Informatics Group
> USDA, National Agricultural Library
> 10301 Baltimore Blvd.
> Beltsville, MD 20705

AAtDB: An *Arabidopsis thaliana* Database

WWW   `http://weeds.mgh.harvard.edu/`
Description   A one-stop shop for links to *Arabidopsis* information. One of the Massachusetts General Hospital servers.

The *Arabidopsis thaliana* Genome Center (ATGC) at the University of Pennsylvania

WWW   `http://cbil.humgen.upenn.edu/~atgc/ATGCUP.html`
Description   Contains physical and genetic map information on *Arabidopsis*.

CottonDB Data Collection Site

WWW   `http://algodon.tamu.edu/`
Description   This site allows data entry for the cotton database, and provides connections to CottonDB at NAL, information about CottonDB, and Email access to the CottonDB researchers.

Maize Genome Database: MaizeDB

WWW   `http://www.agron.missouri.edu/top.html`
Description   Provides the original Sybase version of the maize database as well as links to the gopher and ACeDB versions.

Dendrome : A Genome Database for Forest Trees

WWW   `http://s27w007.pswfs.gov/`

Description   Central repository for information on molecular genetics of forest trees. Nice collection of plant molecular biology protocols. Dendrome Newsletter. TreeGenes gopher server. Image library. Link to WWW version of TreeGenes.

Part II. Genome Information

The Plant Genome Data and Information Center (PGDIC)

WWW   `http://www.nalusda.gov/answers/info_centers/` `pgdic/pgdic.html`

Description   The PGDIC provides in gopher format a calendar of symposia and events, plant genome grant information, a phone book of plant genome researchers, plant DNA library files, plant genome mapping projects, short essays on experimental techniques, the Probe newsletter, full text of biotech patents, and a directory of molecular marker research projects. The PGDIC is a part of the National Agricultural Library, Agricultural Research Service, U.S. Department of Agriculture.

Poplar Molecular Network Gopher

Gopher and FTP `gopher://poplar1.cfr.washington.edu:70/1`

Description   Poplar molecular genetics starting point. Features a newsletter, listserver, RAPD and extraction protocols, genetic maps, and a list of primers.

University of Minnesota Medical School, Computational Biology Centers

WWW   `http://lenti.med.umn.edu/`

Description   The title is deceptive relative to plants, but two gems are contained herein under the heading of research projects: an *Arabidopsis* cDNA sequence analysis; and links to other sequence analysis projects for corn, pine, rice, and others, including a nice set of sequence analysis tools with a tutorial.

Virtual Library of Forest Genetics and Tree Breeding

WWW   `http://www.metla.fi:80/~haapanen/breeding.htm`

Description   Nice links to forest genetics sites worldwide, including sites in Canada, Great Britain, Italy, Finland, Sweden, and the U.S.

Yanofsky Lab's Homepage

WWW   `http://www-biology.ucsd.edu/others/yanofsky/`
              `home.html`
Description Information on the molecular basis of flower development in *Arabidopsis*,
              with special emphasis on the MADS-box genes.

Part III. Seed and Genetic Stock Sources

Nottingham *Arabidopsis* Stock Centre (NASC)

WWW   `http://nasc.nott.ac.uk/`
Description   An on-line *Arabidopsis* seed catalog, order center, andmuch more.
             Browse through over 600 color images. Links to many other *Arabidopsis*
             sites.

***Arabidopsis* Information Management System (AIMS)**

WWW   `http://genesys.cps.msu.edu:3333/`
Description   The AIMS contains a database for *Arabidopsis* seed and DNA stock
             centers worldwide, allowing orders to be made to both the Arabidopsis
             Biological Resource Center (ABRC) at the Ohio State University, and
             the Nottingham *Arabidopsis* Stock Centre (NASC).

LEHLE SEEDS

WWW   `http://www.arabidopsis.com`
Description   Commercial producer of *Arabidopsis* seeds. Catalog of wild type and
             irradiated seeds, as well as growth and seed harvesting systems, and
             books on *Arabidopsis*. Some nice connections are provided to the home
             pages of *Arabidopsis* researchers.

Maize Genetics Cooperation; Stock Center

WWW   `http://w3.ag.uiuc.edu/m aize-coop/mgc-home.html`
Description   The worldwide maize mutant repository. Information and order forms
             for maize genetic stocks are available here.

Part IV. Agricultural Biotechnology

Biotechnology Information Center (BIC)

WWW   `http://www.inform.umd.edu:8080/EdRes/Topic/`
              `AgrEnv/Biotech`

Description   Provides information and links relative to agricultural biotechnology. Of special interest are biotech newsletters and a biotech patent server.

Internet Agricultural Biotechnology Resources

WWW   `http://www.lights.com/gaba/online.html`
Description   This site contains a wealth of links to databases and documents on other servers in the following categories: General Agriculture Resources; Pesticides and Toxic Substances; Biological Controls; University Agriculture Pages; Agricultural Institutes and Societies; Gene and Genome Databases; General Biotechnology Resources; Print Media; Bibliographical Databases and Libraries; Bioreactor information; Food Safety; Commercial Organisations; General Biodiversity Resources; Transgenic Plants and Animals; Biofuels; Hazardous Waste Remediation; Fatty Acids; and Usenet Newsgroups.

NBIAP Agbiotech Online

WWW   `http://gophisb.biochem.vt.edu/`
Description   The National Biological Impact Assessment Program provides documents, news reports, and searchable databases on the impact of agricultural biotechnology. It is administered by Information Systems for Biotechnology, a combined effort of U.S. Department of Agriculture's Cooperative State Research Service and Virginia Polytechnic Institute and State University.

Part V. Molecular Biology Protocols

Molecular Biology Materials and Methods: Indiana University

Gopher   `gopher://iubio.bio.indiana.edu:80/1m/Molecular-Biology/Materials+Methods`
Description   A nice assortment of protocols for common laboratory procedures in molecular biology. This is not limited to plants.

Molecular Biology Protocols

WWW   `http://research.nwfsc.noaa.gov/protocols.html`
Description   A neatly done Web site containing DNA purification, library preparation, and sequencing protocols, PCR methods, electrophoresis information, and links to yet other molecular biology protocol sites.

Part VI. Knowledge Discovery

Knowledge Discovery Mine

WWW   `http://info.gte.com/~kdd/`
Description   Couldn't resist this one. One of the waves of the future in data analysis, presentation and interpretation.

From: *Internet for the Molecular Biologist.*
ISBN 1-898486-02-6 ©1996 Horizon Scientific Press, Wymondham, U.K.

15

MICROBIOLOGICAL RESOURCES ON THE INTERNET

Martin Latterich

Introduction

This chapter contains a compilation of readily accessible internet resources containing material of interest to scientists working with microorganisms. I have attempted to divide this chapter between the more interdisciplinary general resources and the more specialized resources, usually containing information concerning only one model organism. I verified the internet addresses described below, and all were readily available at the time of press. However, many sites change address frequently, and there is no guarantee that they will still be available. It is recommended that in that case a keyword search is made, using one of the search engines or searching the Yahoo database(http://www.yahoo.com/).

There has been an enormous increase in the number of Web sites containing biologically relevant information, and it would be impossible to list all the excellent resources available. The real challenge is to list a representative mix of sites, and in general I chose sites that have many links to other, more specialized sites. I recommend the reader to closely monitor the BIONET newsgroups dealing with microorganisms because very often new Web sites and changes to existing ones are announced there.

Resources of General Interest

American Society for Microbiology (ASM)

WWW	http://www.asmusa.org
Subscriptions	Subscriptions@asmusa.org
Membership	Membership@asmusa.org
Workshops, etc	TrainingInformation@asmusa.org
Career info.	CareerInformation@asmusa.org
Education info.	EducationResources@asmusa.org
Meetings info.	MeetingsInfo@asmusa.org
Description	The American Society for Microbiology has much information for students who are interested in biological careers. It contains information and activities useful to ASM Members, other scientists and the general public.

American Type Culture Collection (ATCC)

WWW   `http://www.atcc.org/`
Description   Catalogues of the American Type Culture Collection (ATCC) can be searched for fungi and other organisms. On-line ordering is available.

Center for Disease Control (CDC)

WWW   `http://www.cdc.gov`
Description   The Center for Disease Control in Atlanta.

dFLASH server

Email   `dflash@watson.ibm.com`
Description   The dFLASH Group has a new electronic mail server that allows GENBANK and PIR similarity searches with the FLASH algorithm.

Notes: For information on this server send the message "help" to: `dflash@watson.ibm.com`. Make sure you have dflash as subject header.

BIONET.Microbiology FAQ

FTP   `mendel.berkeley.edu/pub/micro`
WWW   `http://www.qmw.ac.uk/~rhbm001/BMFaq.html`
Description   The BIOSCI/bionet discussion groups cover many topics including microbiology. See Chapter 9 for more information.

GenBank

WWW   `http://www.ncbi.nlm.nih.gov/`
FTP   `ftp.bio.indiana.edu`; in directory `/Genbank-Sequences/`
Description   A searchable GenBank database. GenBank users can use the World Wide Web (WWW) for submitting sequences to GenBank. The new submission tool, BankIt, provides a simple forms approach for submitting your sequence and descriptive information to GenBank.

GenomeNet

WWW   `http://www.genome.ad.jp/`
Description   GenomeNet is a Japanese computer network for genome research and related research areas in molecular and cellular biology.

Medlab

Email   `LISTSERV@ubvm.cc.buffalo.edu`
Description   MEDLAB-L is a discussion list for medical laboratory professionals. It is run by Patricia Letendre <`pletendr@gpu.srv.ualberta.ca`> in Edmonton. You can join the list by sending "`Subscribe MEDLAB-L`" to the Listserv.

MedWeb

WWW   `http://www.cc.emory.edu/WHSCL medweb.microbiology.html`
Description The MedWeb contains some relevant information of interest to microbiologists.

Microbial Strain Data Network

WWW   `http://www.bdt.org.br/msdn/msdn.html`
Description This server includes a large variety of databases and has links to other microbiologists relevant sites.

Microbiological Underground

WWW   `http://www.ch.ic.ac.uk/medbact/`
Description   This site features an, as yet incomplete, hypertext course on medical bacteriology and contains links to microbiological resources on the net.

Microbiology Virtual Library

WWW   `http://golgi.harvard.edu/biopages/micro.html`
Description   The WWW Virtual Library "Microbiology (Biosciences)", a terrific site with a comprehensive compilation of microbiological links. This resource is the most comprehensive one I have encountered so far. It is a "must" for all readers interested in microbiology and microbiological model systems.

Multi-dimensional Microscopy

WWW   `http://128.205.21.24/`
Description   A Web site containing information on multi-dimensional microscopy.

Society for General Microbiology

WWW   `http://www.mcc.ac.uk/pharmweb/sgm.html`
Description   The Society for General Microbiology have their own Web page
              providing information on the Society, including subscription, and
              forthcoming conferences.

Specialized resources: Viruses

Ebola Information

WWW   `http://www.uct.ac.za/microbiology/`
Description   Information on the Ebola virus.

Hantavirus Home Page

WWW   `http://www.bocklabs.wisc.edu/ed/hanta.html`
Description   Summary of current information on the Hantavirus.

Polio Server

Email   `listserv@sjuvm.stjohns.edu`
Description   To join the Polio List owned by Robert Mauro send an Email message
              `<sub Polio yourfirstname lastname>`.

Virologist Web-page

WWW   `http://www.tulane.edu/~dmsander/`
              `garryfavweb.html`
Description   This page provides a comprehensiv listing of servers for general virology,
              specific viruses, microbiology, AIDS, emerging viruses, electronic
              journals, scientific societies and government sites.

Specialized Resources: Prokaryotes

***E. coli* Database**

FTP   `ftp.embl-heidelberg.de`
Description   Manfred Kroeger's *E. coli* datasets.

E. coli Genetic Stock Center

WWW   http://cgsc.biology.yale.edu/cgsc.html
Email   mary@cgsc.biology.yale.edu
Description   The *E. coli* Genetic Stock Center. This database includes information
              about strains, mutations, genes, and references. For more information
              about the database or to request strains, contact Mary Berlyn at CGSC:
              mary@cgsc.biology.yale.edu

E. coli Index

WWW   http://sun1.bham.ac.uk/bcm4ght6/res.html
Description   The index contains a new section on Microbiology related conferences
              and courses, apart from an excellent section on *E. coli*.

Halophile Directory

Email   simon@uno.cc.geneseo.edu
Description   Contains information on researchers working with halophiles. This list
              will be posted approximately once every two months as a service to the
              community of microbiologists.

Notes: To be added to the list please send Bob Simon an Email message:
simon@uno.cc.geneseo.edu

Mycobacterial Database

WWW   http://www.biochem.kth.se/MycDB.html
Description   Access to this database is available free of charge through the World
              Wide Web.

Tuberculosis Research

WWW   http://molepi.stanford.edu/
Description   A web site for information on tuberculosis research.

Specialized Resources: Eukaryotes

Candida

WWW   http://alces.med.umn.edu/start.html
Gopher   gopher://alces.med.umn.edu:70/11/candida
Description   Information about *Candida albicans* molecular biology. These servers
              share most of their text data. The WWW server has additional images

and diagrams. There are links to the Candida information and other information on the server on that page.

Fungal Genetics Stock Center

WWW   `http://kumchttp.mc.ukans.edu/research/fgsc/`
                `main.html`
                also a UK mirror at SEQNET `http://s-ind2.dl.ac.uk/`
                `FGSC.html`
Description   The Fungal Genetics Stock Center is a resource available to the Fungal Genetics research community and to educational and research organizations in general. The FGSC is funded largely by the National Science Foundation of the United States of America and to a lesser extent by the payments made by researchers who use their services.

Leishmania

Email   `listserv@bdt.ftpt.br`
Description   Mailing list

Notes: To subscribe to the Leishmania list you need to send the following Email:

    Subject:(leave blank)
    Message: `subscribe Leish-L <your full name>`
    Leave the <>brackets out when you type in your name.

Protist Images

WWW   `http://megasun.bch.umontreal.ca/protists/`
                `protists.html`
Description   Protist Image Data provides pictures and short descriptions of selected protist genera, especially those genera whose species are frequently used as experimental organisms or are important in studies of organismal evolution.

Notes: This is a pre-release version.

NIH Campus Yeast Interest Group

WWW   `http://www.nih.gov/sigs/yeast/index.html`
Description   This page contains short descriptions of the research interests of the 12 yeast labs on NIH Campus and some links to other *S. cerevisiae* and *S. pombe* pages.

NIH *S. pombe* Page

WWW   http://www.nih.gov/sigs/yeast/fission.html
Description   This server contains links to several *S. pombe* databases and other web sites.

Pombe Wisdom

WWW   http://t-chappell.mcbl.ucl.ac.uk
Description   The "Pombe Wisdom" site is maintained by Tom Chappell and contains links to databases with *S. pombe* publication abstracts, *S. pombe* genes, vectors and a collection of methods. It contains much important information on pombe.

Notes: Access to the databases requires an ID <pombe> and password <pombe>. This ID may change with time, and you may need to contact Tom for an updated ID and password.

Saccharomyces cerevisiae Genomic Information Resource

WWW   http://genome-www.stanford.edu/
Description   The *Saccharomyces cerevisiae* Genome Center web page. It contains a comprehensive collection of *S.cerevisiae* sequences, genetic and physical maps, and searchable databases. Sequence information from the *S.cerevisiae* genome project will be made available here. The site contains many important links to other *S. cerevisiae* sites.

Schizosaccharomyces pombe Databases

WWW   http://expasy.hcuge.ch/cgi-bin/lists?pombe.txt
FTP   ftp://ftp.sanger.ac.uk/pub/PomBase/PomBase.README
Description   Information about *S. pombe* molecular and cell biology.

The FTP file contains important information concerning a compilation of all *S. pombe* sequences, and how to download and install the *S. pombe* database on computers running the UNIX operating system. A WWW version is in preparation. The WWW resource contains an index of *Schizosaccharomyces pombe* entries and their corresponding gene designations.

Yeasties Home Page

WWW   http://www.sanger.ac.uk/~yeastpub/svw/home.html
Description   The Sanger Center in Cambridge, UK, has a collection on *S. cerevisiae* and *S. pombe* resources.

Glossary

This is a brief glossary of the more common abbreviations and neologisms used in the Internet, and some things which frequently crop up without explanation ("everyone knows that!"). For a very entertaining look at the jargon of computing, both current and historical, see The New Hacker's Dictionary listed in the bibliography.

Address Resolution	The process of turning a domain name into an IP address.
aFTP	Anonymous FTP (q.v.).
Alias	A name other than the official one, used as a convenient mailing address.
Anonymous FTP	The fetching of files from an archive using a user name (usually "anonymous") which is available to all without having to apply for access. Not really anonymous as the password is meant to be your Email address.
Archie	A system for tracking archive contents.
Archive	A collection of data, text or programs, collected together for others to access, typically without charge.
ARPA	Advanced Research Projects Agency, Part of the United States Department of Defence. (Sometimes DARPA, Defence ARPA).
ARPAnet	The precursor to the Internet set up by the ARPA.
ASCII	American Standard Code for Information Interchange. Each letter in a "character set" is assigned an individual and unique number. The ASCII system defines what that number should be for common characters including arabic numerals, alphabetic characters (in both upper and lower case). ASCII is the most common code for representing alphanumeric symbols as numbers in computers. ASCII files contain only alphanumeric characters, as opposed to coded information defining fonts, indices and formatting instructions.
Automagically	Something is jokingly refered to as being done "automagically" if it is performed automatically by a process which is advanced enough to seem magical to all but a few initiates.

Backbone   The part of a network which carries the bulk of the traffic. This will normally be the highest speed part of the network, with multiple, redundant, routes in case of problems.

Bandwidth   The amount of data transfer capacity available.

BBS   Bulletin Board System.

BITNET   "Because It's Time Net" One of the precursors to the Internet, or a part of the Internet, depending on your point of view, and age.

BIX   A computer conferencing system, based in the USA, and a favorite of home computerists. Now accessible by PACNET or Internet, and an access point to many Internet services.

Bounce   Return of undelivered Email: the action, and the error report returned.

btw   By The Way.

CIX   Commercial Interent Exchange.

CFV   "Call for Votes". A formal procedure for voting on the creation of a new Usenet group.

Client   A piece of software which you work with and which obtains information for you by sending requests to a server. The information returned might be modified by the client into a more user-friendly form before presentation to you.

Compuserve   A commercial provider of database and communication services, based in the USA.

Conference   A message area covering a defined subject.

Daemon   A program forming part of the system software, the program is designed to perform a specific function automatically, either after a time interval, or in response to some other event.

Datagram   The basic unit of information or data flow across the Internet. Only a relatively few bytes; lots of datagrams are required for a message or file.

Dial-Up   As a verb, to connect to a computer through the telephone system. As an adjective, describes a port which allows people to so connect (eg a dial-up connection).

DNS   Domain Name System. A distributed database to convert between computer names and numeric addresses, and vice-versa.

Domain A network, or part of a network, under a single administrative umbrella, such as an institution, a group of institutions, or a geographical region.

Domain Name The name of a machine giving its place in a domain, rather than numerically.

Email Electronic mail: this is a word now, not an abbreviation.

FAQ "Frequently Asked Question(s)" Questions which are asked irritatingly often (for example, "What does FAQ mean?"). Often collections of FAQs and their answers are published in an "FAQ List".

Fidonet A network of hobbyist bulletin boards. Fidonet sites often have access to the Internet, at least for Email and News.

Flame An abusive communication, sometimes intended ironically but more often simply rude. Often aimed at people who make mistakes, ask naive or vague questions or who appear out of their depth. Also used as a verb: to flame someone is to attack her in a message, either publicly or privately.

FQDN Fully Qualified Domain Name (q.v.).

FTP From "File Transfer Protocol" The process of, and/or the software used to move files between machines across the Internet.

Fully Qualified Domain Name A machine name with all its heirarchy of domains explicitly given. For example, `ATLAST.CHMEDS.AC.NZ` which under other circumstances might just be called `ATLAST`.

FYI "For Your Information", either as an abbreviation, or as the name of a group of guides to Internet concepts and resources.

Gateway A machine which transfers data between two applications which cannot communicate directly, for example between News and Email, or between two networks which cannot handle each other's data formats, eg between Ethernet and TCP/IP.

Gopher A menu-driven method of accessing Internet resources almost independent of their location.

Host A computer that allows users to communicate with other computers on a network.

HTML Hyper-Text Mark-up Language. Standard method of "marking'" text when constructing World Wide Web documents.

HTTP   Hyper-Text Transfer Protocol.

IMHO   In My Humble Opinion. Most often used ironically, as in "IMHO, the Macintosh is just a toy".

Internet   The grouping of lots of large and small networks which all use the TCP/IP protocol and a common scheme for allocation of machine addresses.

InterNic   Internet Network Information Centre.

IP   Internet Protocol.

ISDN   Integrated Services Digital Network. Single networking system that combines voice and digital communications.

JANET   Joint Academic NETwork. Network connecting educational establishments in the UK.

LAN   A Local Area Network: typically within a building or campus, and usually with one connection to the wider Internet.

Lurker   One who reads a mailing list or newsgroup, but doesn't post.

listserv   Software that manages mailing lists.

Mail Path   The path taken by a mail message to get from the sender to the receiver. Most mail messages will have a long list of machines through which the message has passed.

Mailing List A discussion group communicating by Email.

MIME   Mulit-purpose Internet Mail Extensions, a method of incorporating things other than text (such as graphics, sound, programs) in an Email message for transmission.

Moderator   A person who reviews material being posted to a newsgroup or mailing list for relevance. Only a few groups and lists are moderated.

net.citizen   An inhabitant of cyberspace.

Netiquette   Net Etiquette: the informal, unwritten, unofficial rules for behaviour on the net.

Newbie   Someone who is new to the net, and who may not be up with the folklore and culture.

Newsgroup   A message area on Usenet covering a particular topic area.

NNTP   Network News Transport Protocol

Node   One of the computers connected together in a network.

OPAC   Online Public Access Catalogue.

Packet   A bundle of data.

Port   (1) A physical connection for input or output (eg. a plug on the back of the computer).
                (2) A logical connection number, which distinguishes between many simultaneous connections.

Post   The action of sending a message to a conference, newsgroup etc.

Postmaster The poor sod who is responsible for keeping Email flowing at a site and resolving problems with addresses and the like. Often the network manager; invariably highly talented and under-appreciated. Be kind to a postmaster today!

Protocol   A convention which specifies how two computers will communicate with each other. This will include such low-level details as howbits are transmitted physically and how data packets are identified, and high-level details, such as the way two machines exchange mail messages. Quite dissimilar computers (eg a Macintosh and an IBM mainframe) can communicate using completely different hardware and software, provided that they agree in advance on what the data means.

RFC   Abbreviation for 'Request for Comment'. One of the defining documents for Internet protocols and services.

RFD   "Request for Discussion". A formal request to open a (somewhat ritualised) discussion on the possible creation of a new newsgroup.

Route   The list of machines through which a message is passed between its origin and destination.

Router   A device that directs packets of data between different networks or different parts of a network.

RTFM   "Read the Fantastic Manual". A retort to a question that shouldn't have been asked because the answer is in the manual, or to an incorrect statement made by someone who should know better. The adjective is variable.

Server   Either the software or the hardware part (or both) of a system which offers service(s) to other computers by receiving requests, processing them and returning the results. Examples: Gopher, Archie.

Signal-to-Noise An engineering term, meaning a measurement of the amount of useful information to be found in a larger mass of material. Used humorously, as in "The signal-to-noise ratio in this discussion is zero".

Signature A few lines appended to Email messages and News articles given the sender's name, address and frequently a philosophical quote or witticism. One man's witticism is another's waste of space.

Smiley To avoid a plain text message being misunderstood this system of "Smileys" was developed to signify the writers attitude. For example..

:-(	Sad	:-)	Happy	:-T	Tight Lipped
:-((	Very Sad	:-))	Very Happy	:-v	Speaking
:'-(	Crying	:-X	Kiss	:-V	Shouting
:-ll	Angry	;-)	Winking	:-@	Screaming
:-C	Jaw Hits Floor	<:-)	Dumb Question	:^)	Nose out of joint

SMTP Simple Mail Transfer Protocol: the manner in which mail is transferred between machines, and often by extension used to refer to Email, mailer software etc.

Snail Mail The traditional, terrestrial method of message transmission involving a complex interlpay of pieces of paper, aeroplanes, ships, vans, postmen and dogs!

TCP Transmission Control Protocol.

Telnet The protocol, and the name of the software, used to log on to a remote computer interactively.

UDP User Datagram Protocol.

Unix An operating system used on many large and small computers. More user-hostile than user-friendly in its native form but important even on non-Unix systems as many Internet services use Unix-like commands.

URL Uniform Resource Locator. The Internet address of a resource. Quoted URLs are most often for WWW sites.

Usenet A group of systems that exchange messages posted as part of a newsgroup. This allows global discussion of a vast array of topics.

Veronica A tool that assembles the results of a keyword search into a gopher menu.

WAIS Wide Area Information Service.

WAN   Wide Area Network. cf LAN

Worm   A program which (maliciously) transfers itself from one computer to another, executing on as many computers as possible with the aim of soaking up resources. Not unlike a virus, but not transferred by attaching itself to another program.

WRT   Abbreviation for 'with respect to'.

WWW   World Wide Web.

YMMV   Your Mileage May Vary: "that's what I've found, but your results might not be that good".

Abbreviations or TLAs (Three letter Acronyms)

When reading postings in the newsgroups you will find some acronyms or abbreviations that are used to save typing common phrases in full. These are also called TLAs although the number of letters used is often anything other than three.
Some of the more common are listed here;

AFAICT	As Far As I Can Tell	IMHO	In My Honest/Humble Opinion
AFAIK	As Far As I Know	IMO	In My Opinion
AFK	Away From Keyboard	IOW	In Other Word
AIUI	As I Understand It	JAM	Just A Minute
B4	Before	L8R	Later
BAK	Back At Keyboard	LOL	Laughs Out Loud
BBL	Be Back Later	NALOPKT	Not A Lot Of People Know That
BCNU	Be Seeing You	NRN	No Reply Necessary
BSF	But Seriously Folks	OEM	Original Equipment Manufacturer
BTDT	Been There Done It	OTOH	On The Other Hand
BTW	By The Way	PD	Public Domain
DWISNWID	Do What I Say Not What I Do	ROFL	Rolls On Floor Laughing
EOF	End Of File	RTFAQ	Read The FAQ
F2F	Face to Face	RUOK	Are You Okay?
FAQ	Frequently Asked Questions	TIA	Thanks In Advance
FOC	Free Of Charge	TNX	Thanks
FWIW	For What Its Worth	TTFN	Ta Ta For Now
FYA	For Your Amusement	TTYL	Talk To You Later
FYE	For Your Entertainment	TVM	Thanks Very Much
FYI	For Your information	WRT	With Regard To
<G>	Grin!	WYSIWYG	What You See Is What You Get
GA	Go Ahead		
IAE	In Any Event		
IME	In My Experience		

Index

ORNL 140
OWL 128

P

PGDIC 165
PIR 62, 131
Plant Molecular Biology 163-169
Polio 172
Pombe 175
Poplar 165
Postmaster 24
PPP 49
ProDom 135
PromoterScan 96
PROSITE 80, 135
Protist 174
Protocols 167-168
Publishing 55
Pythia 74, 137

Q

QUICKSEARCH 66

R

Rat 146
REBASE 135
Reference Library Database 144
Repeats (sequence) 74
Request for Comment 12
Restriction Enzyme DataBase 135
Retrieve 62
RLDB 144
Rodent-Research 126

S

Saccharomyces 154, 175
SASE 136
Schizosaccharomyces 175
Seeds 166
Sequence Databases 129
Sequence Retrieval System 128
SGM 172
SIGMA 140
Signatures 25
Sl/IP 49
Social implications 138
Society for General Microbiology 172

SORFIND 90
SRS 128
SWISS-PROT 76, 131

T

TBASE 134
TELNET 32, 49
TFD 136
Tools 146
Transcription Factors 136-7
TRANSFAC 137
Transgenic Animals 134
Transmission Control Protocol/Internet
    Protocol, TCP/IP 11, 49
Tuberculosis 173
Tumor Gene Database 134

U

Universal Resource Locator, URL 50
Usenet News 29, 101-105, 108

W

WEB browsers 48
Whitehead centre 140
Wide Area Information System, WAIS 37
World Wide Web, WWW 37

X

Xblast 75
Xpound 94

Y

Yeast 155, 174